T0260191

DIGITAL CIRCUIT BOARDS

DIGITAL CIRCUIT BOARDS
Mach 1 GHz

Ralph Morrison

A JOHN WILEY & SONS, INC., PUBLICATION

For general information on our other products and services or for technical support, please contact our Customer Care Department within the United States at (800) 762–2974, outside the United States at (317) 572–3993 or fax (317) 572–4002.

Wiley also publishes its books in a variety of electronic formats. Some content that appears in print may not be available in electronic formats. For more information about Wiley products, visit our web site at www.wiley.com.

Library of Congress Cataloging-in-Publication Data:

Morrison, Ralph.
  Digital circuit boards : mach 1 ghz / Ralph Morrison.
      p. cm.
  Includes bibliographical references.
  ISBN 978-1-118-23532-4
1. Digital electronics. 2. Logic design. 3. Integrated circuits. I. Title.
  TK7868.D5M68 2012
  621.382–dc23

                                                2011043534

10 9 8 7 6 5 4 3 2 1

Buildings have walls and halls.
People travel in the halls not the walls.
Circuits have traces and spaces.
Energy travels in the spaces not the traces.
Ralph Morrison

A word about the book title

Mach 1 was a barrier in flight for a long time. Aircraft that can go faster than the speed of sound are more expensive and more difficult to design. There is a barrier in digital design that occurs at clock rates around 1 GHz. One clock period is one nanosecond, and in this time an electromagnetic wave can travel about 15 cm in epoxy. This is the dimension of a typical circuit board. Circuits that can perform near or above 1 GHz present a new set of design challenges. In that sense, there is a barrier to cross. This book discusses the challenges of designing these faster and faster circuits. Many old ideas must be discarded and new ones accepted. There are no sonic booms that I know of. I hope the ride through Mach 1 is a smooth one.

CONTENTS

Preface   **xi**

1 BASICS   1

1.1   Introduction   1
1.2   Why the Field Approach is Important   3
1.3   The Role of Circuit Analysis   4
1.4   Getting Started   5
1.5   Voltage and the Electric Field   6
1.6   Current   7
1.7   Capacitance   8
1.8   Mutual and Self-Capacitance   10
1.9 E Fields Inside Conductors   11
1.10 The D Field   12
1.11 Energy Storage in a Capacitor   12
1.12 The Energy Stored in an Electric Field   13
1.13 The Magnetic Field   13
1.14 Rise Time/Fall Time   15
1.15 Moving Energy into Components   15
1.16 Faraday's Law   16
1.17 Self- and Mutual Inductance   16
1.18 Poynting's Vector   17
1.19 Fields at DC   18
Glossary   19

2 TRANSMISSION LINES   22

2.1   Introduction   22
2.2   Some Common Assumptions   24
2.3   Transmission Line Types   25
2.4   Characteristic Impedance   27

2.5	Wave Velocity	29
2.6	Step Waves on a Properly Terminated Line	30
2.7	The Open Circuited Transmission Line	31
2.8	The Short Circuited Transmission Line	33
2.9	Waves that Transition between Lines with Different Characteristic Impedances	35
2.10	Nonlinear Terminations	38
2.11	Discharging a Charged Open Transmission Line	38
2.12	Ground/Power Planes	40
2.13	The Ground and Power Planes as a Tapered Transmission Line	41
2.14	Pulling Energy from a Tapered Transmission Line (TTL)	43
2.15	The Energy Flow Through Cascaded (Series) Transmission Lines	45
2.16	An Analysis of Cascaded Transmission Lines	48
2.17	Series (Source) Terminating a Transmission Line	49
2.18	Parallel (Shunt) Terminations	50
2.19	Stubs	52
2.20	Decoupling Capacitor as a Stub	54
2.21	Transmission Line Networks	54
2.22	The Network Program	55
2.23	Measuring Characteristic Impedance	56
	Glossary	57

3 RADIATION AND INTERFERENCE COUPLING — **61**

3.1	Introduction	61
3.2	The Nature of Fields in Logic Structures	62
3.3	Classical Radiation	62
3.4	Radiation from Step Function Waves	63
3.5	Common Mode and Normal Mode	66
3.6	The Radiation Pattern along a Transmission Line	70
3.7	Notes on Radiation	70
3.8	The Cross Coupling Process (Cross Talk)	71
3.9	Magnetic Component of Cross Coupling	72
3.10	Capacitive Component of Cross Coupling	74
3.11	Cross Coupling Continued	75
3.12	Cross Coupling between Parallel Transmission Lines of Equal Length	76
3.13	Radiation from Board Edges	78
3.14	Ground Bounce	79

3.15   Susceptibility   80
Glossary   80

4 ENERGY MANAGEMENT   82

4.1   Introduction   82
4.2   The Power Time Constant   84
4.3   Capacitors   86
4.4   The Four-Terminal Capacitor or DTL   87
4.5   Types of DTLs   89
4.6   Circuit Board Resonances   90
4.7   Decoupling Capacitors   90
4.8   The Board Decoupling Problem   92
4.9   The IC Decoupling Problem   93
4.10   Comments on Energy Management   94
4.11   Skin Effect   95
4.12   Dielectric Losses   97
4.13   Split Ground/Power Planes   97
4.14   The Analog/digital Interface Problem   98
4.15   Power Dissipation   99
4.16   Traces through Conducting Planes   100
4.17   Trace Geometries that Reduce Termination Resistor Counts   101
4.18   The Control of Connecting Spaces   101
4.19   Another way to look at Energy Flow in Transmission Lines   103
Glossary   104

5 SIGNAL INTEGRITY ENGINEERING   106

5.1   Introduction   106
5.2   The Envelope of Permitted Logic Levels   107
5.3   Net Lists   108
5.4   Noise Budgets   108
5.5   Logic Level Variation   109
5.6   Logic and Voltage Drops   110
5.7   Measuring the Performance of a Net   111
5.8   The Decoupling Capacitor   112
5.9   Cross Coupling Problems   114
5.10   Characteristic Impedance and the Error Budget   114
5.11   Resistor Networks   116
5.12   Ferrite Beads   117

5.13   Grounding in Facilities: A Brief Review   118
5.14   Grounding as Applied to Electronic Hardware   120
5.15   Internal Grounding of a Digital Circuit Board   123
5.16   Power Line Interference   124
5.17   Electrostatic Discharge   125
Glossary   126

6 CIRCUIT BOARDS   **130**
6.1   Introduction   130
6.2   More about Characteristic Impedance   131
6.3   Microstrip   133
6.4   Centered Stripline   135
6.5   Embedded Microstrip   136
6.6   Asymmetric Stripline   137
6.7   Two-Layer Boards   140
6.8   Four-Layer Circuit Board   143
6.9   Six-Layer Boards   145
Glossary   147

Abbreviations and Acronyms   **149**
Bibliography   **157**
Index   **159**

PREFACE

The story of Digital Circuit Boards at Mach 1 GHz starts with my friend Daniel Beeker. Dan is a senior field applications engineer for Freescale Semiconductor. He was instrumental in getting me interested in circuit board design problems. He was the one that spurred me into finishing the 5th edition to my book "Grounding and Shielding," which was published by John Wiley in 2007. I have rewritten this book five times since 1967 and when this fifth writing was finished, I really thought I was through writing books. Obviously, I was mistaken.

Dan sees the problems encountered by his customers. He recommended to his management that the users must be provided some help in the form of seminars. They agreed, and as a result Dan took on a new set of responsibilities. He was tasked to find speakers and arrange for seminars for Freescale customers. To locate speakers, Dan turned to his own personal library. The first book he took from the shelf was a copy of my "Grounding and Shielding." He then looked into my web site and found my address. The result was that I was invited to participate in the Freescale Forum in Orlando and later to give a seminar to his customers in the Detroit area.

The seminar I gave was based on my book and was well received. For those familiar with my books, I use very simple physics to explain how interference is generated, how it enters circuits, and how the circuits can be protected. The principles are the same whether the problem is analog, digital, or rf. I had little trouble bringing transmission line theory into my discussions. I found I had to catch up on the language of circuit boards and how they are built. I had to find out what a BGA was, what prepreg meant, and what is an interposer board. I had to learn the difference between a blind and a buried via.

Fortunately, Dan followed through with additional seminars where the speakers understood the details of circuit board design and could relate more closely with the details of components and materials. I admire Dan for recognizing that the fundamentals must come first. Even though I knew little about the details of board design, I could show the users how and why layout geometry was critical if they were going to build successful boards.

Dan wanted me to get closer to the circuit board problems, so he arranged for me to attend the PCB conference in Santa Clara, given by the UP Media Group. At the conference I sat in on courses given by experts in the field. I learned a lot about how circuit boards were designed and built. The talks introduced me to the designers' problems. Many of the speakers used a combination of circuit theory and lore to explain circuit board behavior. This is just the problem I had been dealing with throughout my career. I was in a different field with a new language. I had a lot to learn. I wanted to understand the digital layout problem based on physics, not on lore.

I found that digital engineers were working in areas of nanosecond delays and picosecond rise times. This was an area where I understood the physics but not the details of board construction. I recognized that real time delays were involved, and this was not covered by circuit theory. The speakers related their real world experiences and how they resolved many difficult problems. I remember speakers saying that energy could be drawn from the ground/power plane faster than from a capacitor. This got the wheels turning. How fast is fast and how much energy is there? How fast are capacitors?

I knew the basic physics so the challenge was to learn the language and make sense out of all the material that was being presented. When I got home from the show one of the first things I examined was how energy is moved from a ground/power plane. I assumed a coaxial connection to the conducting planes. If a step load was placed on this connection, the wave that propagated outward moved in a circular pattern. The characteristic impedance of the wave depended on the radial distance from the point of coaxial connection. I recognized that there were continuous reflections as the wave propagated outward and that the energy returned to the source increased with time. Much to my surprise, I found that the power curve depended on conductor spacing and not on the dielectric constant. With a greater dielectric constant, the energy was pulled from a smaller area. With a higher dielectric constant and with multiple demands for energy there would be less cross talk.

This exercise helped me understand the problem of connecting to the ground/power plane. I recognized that vias were the accepted method of making connections between layers. Typically, a via geometry has an aspect ratio of unity, where the characteristic impedance is about 50 ohm. When I assumed that a short lead length of 50 ohm was used to connect the ground/power plane to a load, I found out a very important fact. A short section of coaxial line placed between the load and the ground/power plane increased the rise time significantly. What was happening was that a large number of wave reflections were required to move energy across this short connection.

This one fact really caught my attention. It had a far reaching impact on my thinking. How many places were there, where short sections of transmission lines were used to connect a load to a source of energy?

The first area I considered was the capacitor. The available books and all the speakers discussed the natural frequencies of capacitors. The limitations in performance were assigned to the series inductance. This was a good explanation for axial lead capacitors but what about surface mounted components? Because capacitors of any geometry have a natural frequency when tested with sine waves, the assumption that resonance is related to a simple series inductance is made. In my view, I was dealing with step functions and reflections, and these ideas of circuit theory did not exactly fit. I decided then that I needed to look at digital processes in a consistent manner. I could not mix step function discussions with sine wave terminology. I had to clean up my understanding and explain things in an unambiguous manner. Resonant frequency concepts using sine waves was not compatible with real time reflections on a transmission line.

Calling a capacitor a short transmission line was the first step. Then I realized that the symbol was misleading as it implied a midpoint connection. Then some obvious issues came to mind. How can you take energy out of a capacitor and put it back in at the same time? What kind of construction allows wave energy to enter between the conductors? If you want energy with a short time constant, how do you construct a low impedance connection to a capacitor? In working on these problems, a book was beginning to form.

I have been doing engineering and consulting while writing books for some time. I have come to the conclusion that we often use electrical symbols in a careless manner. We need the symbols, but they represent complex conductor geometries that are intended to store or dissipate field energy. This is the correct view for all electrical activity whether we describe circuit behavior in terms of step functions or sine waves. Unfortunately, the field view is cumbersome, difficult mathematically, and impractical most of the time. Circuit theory is always correct, but it requires that the simplifications that are made are applicable. For example, simple circuit theory does not allow for time delays and that eliminates transmission lines. It implies that the interconnecting leads do not affect circuit performance, as we assume zero resistance and inductance for conductors regardless of length. We can always include parasitic elements in any analysis, but when we need them they are still just approximations. A resistor at 1 MHz can be represented as having a single parasitic shunt capacitance. At 100 MHz, the lead inductance starts to play a role and a distributed parasitic capacitance is needed. At 1 GHz, the resistor becomes a part of a transmission line path and the return path geometry is critical. At 10 GHz, there is really no such thing as a resistor even if we add many correcting terms. It is a lossy conductor geometry that modifies an electromagnetic field. The problem we face is how do we represent this device? The symbols we draw are very misleading. So far, there are no suggestions on what else we might do.

In my "Grounding and Shielding book," I recognized the need to use field theory to explain electrical activity. I kept it very nonmathematical. The ideas are important and not the exact numbers. This book is no different. The basic physics is again the starting point. In this book, the emphasis is on events that occur in picoseconds where the signals are in the volt range. In "Grounding and Shielding," the problems were usually related to microvolt signals, where the frequencies of interest were under 1 MHz. Move the signal level and signal bandwidth up to six orders of magnitude to volts and gigahertz, and the same physics solves a different set of problems.

I treat the basics in Chapter 1. It is important for the reader to appreciate that all the answers to their problems rest somewhere in this Chapter 1. If the reader needs a more in-depth review, I suggest reading the first chapters of "Grounding and Shielding" book. Of course, there is always that textbook from school that rests on the nearby bookshelf.

This book is intended to help the reader understand the problems of laying out digital circuit boards for fast logic. It is not intended as a treatise on how circuit boards are made although this knowledge can often be very helpful. Understanding

xiv   PREFACE

the manufacturing process helps in understanding practical design. An engineer needs to know the range of trace widths that can be accommodated or how thin a dielectric can be used or what materials will withstand soldering. He needs to understand the cost of different laminates and why they are needed. All of this comes from the experience gained in doing designs and working with a manufacturer. On top of all this knowledge, the engineer needs to know the basics of signal transmission so that the logic will function.

Learning is an ongoing process. Board manufacturers will continue to improve their art. The only constant thing will be physics. It was my intent in writing this book to stick with the basics and use the present art as an example when it seems relevant. It is interesting to note that the materials we use today are the same ones that were used 20 years ago. Change is usually a refinement in processing raw materials caused by the continuous demand to improve performance and reduce cost. I hope this book will make a few design tasks a bit easier.

RALPH MORRISON
Pacifica, CA

April 2012

1

BASICS

1.1 INTRODUCTION

This book is written to provide the reader with the basic understanding that is needed to layout high speed digital circuit boards. The wiring and circuitry that interconnects components mounted on these boards has become a new field of engineering. This book treats the analog aspects of digital board design, as it relates to the hardware that is selected. The term *analog* applies to such topics such as rise and fall time, overshoot, transmission delay, reflections, radiation, settling time, cross talk, and energy flow. The term *analog* is often used to describe circuits that use a carrier signal, such as in cell phone communication. Much of the material in this book will apply to carrier signals, but the emphasis will be on digital circuit board layout. The methods described in this book work well for slower logic. These methods do not add to board cost. These methods do add to reliability and performance. Software design and selection of components that make up a logic design are not discussed.

Rather than gather a big list at the end of the book, I have placed a limited glossary at the end of each chapter. I feel that this glossary should contain the words that are most critical to an understanding of the material in the text. In an industry that is changing very rapidly, there are bound to be language problems. This is the

Digital Circuit Boards: Mach 1 GHz, First Edition. Ralph Morrison.
© 2012 John Wiley & Sons, Inc. Published 2012 by John Wiley & Sons, Inc.

case with the circuit board industry and digital logic. The definitions that are used must be as clear as possible or the reader will not be helped.

The reader is encouraged to read over the glossary after completing each chapter. It will serve as both a review and a test of understanding. If a phrase or expression is not found then the next step is to use the index. If I have left something out, the internet can provide some assistance.

Electrical phenomena can be explained using basic physics, but the words that are used must be carefully chosen. Some phrases have a way of changing meaning over time. The term *critical length* was first used by transmission line engineers in the early days of radio where the signals were sine waves. Today, the expression is applied to cross coupling between traces on a circuit board where digital logic is involved. By reapplying the term, some of the original meaning has changed.

Abbreviations are not listed or used in the glossary of terms. Only well-accepted abbreviations and acronyms are used in the text. The reader is referred to a list at the end of this book.

In describing electrical behavior, the explanations vary with frequency and with specific disciplines. At clock rates above 100 MHz, many of the concepts used in circuit analysis begin to fail. In circuit analysis, there is a time delay associated with phase shift. The steady state operation of a circuit may occur after a hundred sine wave cycles have gone by. This implies that in a circuit, there are transient effects that must totally attenuate before the steady state solution is available. In digital circuits, there is a time delay associated with signal propagation. There is no waiting for one hundred cycles while transients decay. Delays in analog circuitry must be treated quite differently than the delays caused by signals traveling on a transmission line.

In a digital circuit, every connecting trace and every component on a circuit board can be considered a transmission line. It takes time for a signal to travel the length of a transmission line, and this time is independent of clock rate. It takes time to obtain energy from a capacitor. In circuit analysis, there is no simple way to treat these delays or the propagation delay of a transmission line.

In a resonant circuit involving inductance and capacitance, the energy flows back and forth between the two components at one frequency. The analysis is called *steady state* when all transient effects have attenuated and the performance repeats for each cycle of the driving signal. When a wave reflects back and forth on a transmission line, there is a time of transit not found in a resonant circuit. There is a spectrum associated with every logic transmission that relates to the rise time and extends toward dc. Circuits are analyzed one frequency at a time. Logic reacts to signals that are composites of an entire spectrum of sine waves. Thus, the character of a logic signal is very different than the character of a sine wave, yet the term *frequency* is used for both.

Printed circuit boards (PCBs) or printed wiring boards started out as a way to avoid hand soldering the interconnection of electronic components. Early boards had traces on one or two sides of an epoxy board with few plated-through holes. As component densities and clock speeds increased, it became necessary to include conducting planes in the design. These planes are used to distribute power and ground to components and to provide a return path for signal currents. Today, many boards are manufactured with dozens of interconnected layers, with many ground and power planes using surface-mounted and embedded components. As the clock rates have risen, board designs have become more and more of an engineering problem. It is no longer an issue of simply connecting the components together. A few of the problem areas that we will discuss include traces that jump between layers, the spacing of vias, trace routing that uses stubs, energy distribution, energy dissipation, board radiation, and signal cross talk.

Today's circuit board designs require engineering, and this engineering must be based on physics. This physics controls the details of design and is the basis of this book. Effective engineering must consider price and performance, as well as topics such as radiation and susceptibility. Surprisingly, there is a lot to say.

Circuits are often thought of as a configuration of components. When the logic rise times are associated with 1 MHz clock rates, this is basically true. The leads that connect the circuit can be routed almost at random, and there will be few problems. In low signal level analog designs below 100 kHz, lead routing can make or break the product. In digital circuits operating at clock rates above 100 MHz, the routing of all leads is as important as the selection of components. This book discusses the engineering of board layout and wiring, so that logic boards can function correctly; and at the same time, the engineer can control cost and limit radiation.

N.B.

The rise and fall time of logic signals is more critical than the clock rate.

1.2 WHY THE FIELD APPROACH IS IMPORTANT

All electrical behavior from dc to light can be described in terms of the electric and magnetic fields. For many reasons, a field approach to circuit function is very impractical. This is why engineers heavily rely on circuit theory as a working tool. Our understanding of how a circuit functions is closely related to the circuit symbols we have created and to the language we use. When we use capacitors and inductors in our analysis we think of reactances and we generally ignore the fields and energy storage that are inside of these components. In a typical circuit design, the energy that is moved and stored between traces or between traces and a conducting plane is not considered. In high speed logic, this movement of energy must be considered. In fact, this energy must be controlled and dissipated so that the logic can function. The dissipation of this energy can be a serious problem as it can cause board overheating.

Every component is a conductor geometry of some sort. Fields inside the components determine their performance. In a FET (field-effect transistor), there is an electric field between the source and drain. Fields carry operating power and signals to the components over the connecting traces. Getting these fields to the components on a timely basis is handled by traces. The routing of traces is a problem, and we discuss this in great detail later.

Nature does not read our circuit diagrams or symbols. She approaches a circuit as a conductor geometry that allows the flow of electromagnetic energy. Her one goal is to find a way to store less field energy. We use this one fact to get her to perform electrical tasks. A design usually starts by providing a power supply that is a source of energy. Energy leaves this power source as voltage and current. The energy actually flows in electromagnetic fields that follow in the spaces between conductor pairs. These pathways spread the energy, which leads to various losses. Consider the parallel with the flow of water. A dam stores water that we allow to flow in conduits. These conduits reach our homes to supply water for many uses. In the city, we channel water in storm drains to limit water damage during a storm. These same ideas of flow apply to our circuits. If we cooperate with nature, we can make effective use of this energy flow. Our designs must keep the energy from a storm (radiation) from entering our circuits. We want to control energy paths, so the energy required in one circuit does not interfere with another circuit.

At a sufficiently high frequency, components lose their simple circuit identity. A capacitor, for example, can be viewed as having a series resistance and inductance. The inductance implies that a magnetic field is involved in energy transport. We will see that a simple circuit theory approach may not be effective. If we turn to field theory for help, we will not get exact answers. If we want to appreciate what is actually happening in a circuit, both field theory and circuit theory must be applied to the problem. We must learn how to use the field approach so that we can make good engineering decisions.

1.3   THE ROLE OF CIRCUIT ANALYSIS

We are used to thinking in terms of sinusoids. The language of circuit analysis is closely related to these sine wave signals. The words *impedance* or *reactance* are defined in terms of sine waves. We are married to circuit symbols and sinusoidal measures and somehow we must stretch these meanings to fit a high speed digital world. We certainly are not going to invent a new language or use new symbols to describe our needs. The old words and symbols will have to do. Usage defines meaning and eventually usage may change. We will use the word *impedance* without regard to sinusoids.

In digital circuitry, the signals are usually in the form of step functions of voltage. These voltage level changes are applied to traces that carry information (energy) between components. We will refer to these traces as transmission lines. It will come as a surprise to some readers that a 1/16-in-long trace should be considered a transmission line. Also that a vias associated with a trace can modify the character of the transmission path. The reflections on sections of the energy path can result in radiation and limit performance.

Repetitive wave forms are equivalent to a group of sine waves with a harmonic relationship. For example, a square wave can be equated to a group of sine waves consisting of a fundamental and all the odd harmonics of this frequency. The amplitude of each harmonic is inversely proportional to the number of the harmonic. Digital signals are not repetitive and a harmonic wave analysis may not be directly applicable. Some of the ideas expressed by harmonic analysis can still be of value. The most important factor that we must consider is the rise and fall time of the leading edges in step functions. Sometimes, the third harmonic of the fundamental (clock frequency) can be used as a reference frequency in describing or analyzing some aspect of performance. This makes sense if the rise and fall times are 20% of the clock period.

The materials used in board manufacture have been around for a long time. Progress is the process of using these same materials in new and better configurations to serve our needs. A lot of engineering is based on past practice. Every practice has its limitations. We must question everything so that when a practice needs changing we can act accordingly. A better design will most likely be a change in how we configure these available materials. Some of the material is given to us in the form of components and some of the material is used in mounting and connecting these components. Our job is to select the right components and configure them on circuit boards so that field energy flows to perform the task at hand.

1.4   GETTING STARTED

This book assumes that the reader is familiar with the basic physics of electricity and with basic circuit theory. There are a few ideas that need to be stressed before we discuss circuit board design. A review is a good idea and I hope that the reader will take the time to read these few paragraphs.[1] The intent is to stress concepts. Simple equations will be presented because they are the clearest way to state relationships.

In an isolated uncharged conductor there is a balance of charge. In every atom, the inner protons carry a positive charge equal to the negative charge on the electrons. This balance is extremely precise. In a conductor, electrons can easily move between atoms. We call this *motion current*. The percentage of available electrons that partake in electrical circuit activity is so small that it is hard to describe it in useful terms. The ratio is like that of a teaspoon of sand to miles of beach.

The forces that exist between electrons are not easily imagined. Consider that 1% of the electrons in a human being are free to interact with the same number of electrons on a second person. The force between the charges would be enough to lift the earth out of orbit. It is obvious from this fact that the number of electrons involved in any electrical activity in our circuits is indeed very small. It is hard to realize that this immense electrical force is involved in current flow.

The protons are so well shielded inside the atom that they do not partake in normal circuit activity. In the circuits we will consider, most of the electrons move on the surface of conductors. To start, consider a group of extra electrons placed on a conductor. These electrons are free to move between atoms. The forces between these

---

[1] A more thorough but elementary treatment of these fields is given in the author's book "Grounding and Shielding—Circuits and Interference" 5th Edition, John Wiley, publisher.

"free" electrons cause them to move apart as far as they can go. On a conductor, they assume a position that stores the least amount of potential energy. On an isolated sphere, they end up spaced uniformly on the entire outer surface.

An accumulation of added electrons is called a *negative charge*. A depletion of electrons from a conductor is called a *positive charge*. These pseudopositive charges attract electrons and move just like electrons. When a pseudopositive charge meets an electron, they are canceled. In other words, the electron is accepted by an atom to fill in its outer shell. In most of our discussions we will discuss the motion of charges and the direction of current flow. The polarity of surface charges will not be discussed.

1.5   VOLTAGE AND THE ELECTRIC FIELD

The force field that exists around any group of charges is called the E or *electric field*. It is a vector field, as it has intensity and direction at all points in space. The charges (or absence of charge) we will consider are usually distributed on the surfaces of circuit conductors. This force field in the space around charged objects can be sensed by placing a very small test charge in the field. The test charge has to be small enough that it does not contribute to the field being tested. Thus, a test charge is a small accumulation of charge on a small mass.

Work must be done in moving this test charge in an electric field. The work required to move a unit charge between two points is called the *potential difference* (voltage) between those two points. We usually measure potential difference between conductors. In a radiated field there are no conductors to consider, yet there are potential differences.

Definition: *Voltage* is the work required to move a unit charge over a distance in an electric force field.

If there are voltage differences, there are electric fields. The converse is also true. If there are electric fields there must be voltage differences. If there are charges on a surface there must be an electric field. Conversely, if there are nonradiated electric fields there must be charges on conductive surfaces.

Voltage differences can exist between points in space or between conducting surfaces. Electric fields exist at all frequencies including dc. Electric fields are represented by curves that follow the direction of the field forces. In a field representation, lines of force start on a fixed amount of positive charge and terminate on the same amount of opposite charge. When the lines are close together, the forces are the greatest. In a field representation it is only necessary to use a limited number of lines to outline the shape of the field. When there is voltage, electric field lines terminate on the surface charges of a conductor. As we will see a very small electric field inside, a conductor is required for current flow. When there is no current flow there is no field inside of a conductor.

The words *ground* or *ground plane* will be used frequently. A ground is a conducting surface that is larger than the surrounding circuitry. In a facility, the earth may be called *ground*. In an integrated circuit die, a conducting surface can be called *ground*. On a circuit board, a conducting plane can be called *ground*. A ground has the quality that it allows charges to move freely on the surface and collect where the field line terminates. An ideal ground plane has no potential differences from point to point. In most circuits this is very nearly true. A superconductor can have current flow with zero internal electric field.

1.6 CURRENT

The charges that create a static electric field are located on the surface of conductors. In every circuit with operating voltages, there are surface charges. These surface charges must move if the voltages change value. The pattern of charge motion can best be appreciated by noting the pattern of electric field lines that terminate on a conductor. Charges tend to concentrate where there are sharp bends or transitions. This means that if the field pattern changes, current flow must occur when the pattern changes. If there are no field lines terminating on a part of a conductor then there will be no current flow on that part of the conductor. This means that parts of a ground plane are often not used for current flow. Figure 1.1 shows the E field pattern around a typical trace over a ground plane.

The field pattern shows that the charges concentrate on one side of the trace and on that portion of the ground plane under the trace. No current flows on the other side of the ground plane. This charge pattern is essentially independent of frequency. Every E field line terminates on a unit of charge. The unit of charge is selected to best show the shape of the field. Note that the electric field lines terminate perpendicular to conducting surfaces. The lines that are shown parallel to the ground plane represent equipotential surfaces. The trace (conductor) itself is an equipotential surface at voltage V. The ground plane is an equipotential surface at zero volts. The dielectric normally present under a trace is not shown. A dielectric would not significantly change the field pattern, but it would increase the amount of stored charge.

The field pattern is shown for a voltage V. As V is increased, the number of lines also increases. If the voltage is a sine wave the field intensity pattern is also sinusoidal.

If there is sinusoidal current flow below a few kilohertz, a very small horizontal component of the E field penetrates into the conductors to move the charges. As the

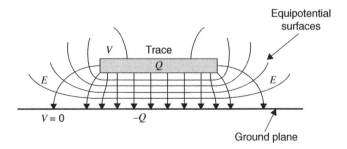

Figure 1.1 The electric field pattern around a typical circuit trace.

frequency increases, the penetration depth decreases and current flow concentrates closer to the surface. Above a megahertz, the current flow in conductors is basically at the surface. This surface flow phenomena is called *skin effect* (Section 1.8).

N.B.

The electric field pattern under a trace primarily depends on the instantaneous voltage between the trace and the ground plane. It is not dependent on current level, clock rate, or sinusoidal frequency.

The velocity of electrons in a typical circuit is extremely low. In a typical geometry, it is not too difficult to show that the average velocity is less than 1 cm/s. This occurs because the electron density in a conductor is extremely large. What is important is the nature of the electric and magnetic fields. Fields carry signals and energy at the speed of light. For this reason, we must concentrate on the movement of fields and not on the flow of electrons. We need to know the value of current flow to calculate voltage drop, power losses, and magnetic field strength. This is a good example of where circuit theory can be put to good use.

N.B.

When there are potential differences in a circuit, there are surface charges present wherever E fields terminate. These charges are the ones that move when current flows. It takes a very small E field component inside the conductor to move these electrons.

1.7 CAPACITANCE

When a charge is moved to a new position in an electric field, the work that is done is stored in the electric field. If the electric field is confined to a region between two conductors, the conductor geometry is called a *capacitor*. The capacitance of this geometry is the ratio of stored charge to potential difference.

N.B.

Capacitance is the ability to store electric field energy. A capacitor is a conductor geometry that is designed to store electric field energy. The energy is stored in space and not in the conductors.

A volume in space with an electric field stores field energy. This volume has capacitance, although it may not be in a capacitor. It is important to appreciate that nearly all field energy is stored in space. (Remember that in most practical circuits it takes very little E field in the conductors to cause current to flow. This means that there is very little field energy stored in the conductors.) Our hope is that we can confine this energy to very small regions under and between traces and in capacitors. This limits the amount of "loose" energy involved. This is how we control radiation. Conductor geometry defines the shape and extent of the field. If it is confined by conductors, it cannot radiate.

N.B.

At low frequencies (below 100 kHz) most of the field energy generated by a circuit returns to that circuit.

The intensity of the electric field is reduced by the presence of a dielectric. If the charge in a capacitor is held constant and a dielectric is inserted, the voltage must drop. The ratio of charge to voltage (capacitance) thus increases by a factor equal to the relative dielectric constant. The capacitance of two parallel plates with an area A and a spacing h is equal to

$$C = \frac{A\varepsilon_R\varepsilon_0}{h} \tag{1.1}$$

where ε_0 is the permittivity of free space equal to 8.85×10^{-12} F/m, ε_R is the relative dielectric constant, A is the plate area, and h is the spacing. The unit of capacitance is the farad. This equation assumes that there is no fringing of the field at the edges of the plates. This is a good assumption when the spacing between conductors is very small.

Equation 1.2 states that the ratio of stored charge to voltage is a measure of capacitance.

$$C = \frac{Q}{V} \tag{1.2}$$

If the stored charge is one coulomb for a voltage of 1 V then the capacitance is 1 F. In most circuit applications a farad is a very large unit. Typical decoupling capacitors used on a circuit board range from 1000 to 10,000 pF (pF stands for picofarad or 10^{-12} F).

A fundamental relation exists between current flow, voltage, and capacitance. If the voltage rises at 1 V/s, the current flow in a 1-F capacitor is 1 A. For a capacitor of 1 μF, if the voltage rises at 1 V/s, the current flow is 1 μA. This fact is expressed in Equation 1.3 in terms of the derivative of voltage with respect to time.

$$I = C\frac{dV}{dt} \tag{1.3}$$

1.8 MUTUAL AND SELF-CAPACITANCE

In a typical capacitor, the electric field lines start on one conductor and terminate on the second conductor. The conductor at zero volts is often called the *reference conductor*. The ratio of charge on the first conductor to the charge on that of the second conductor is called *self-capacitance*. The capacitors used in a circuit all have self-capacitances.

When there is a group of conductors as in a circuit, the concept of capacitance must be extended to allow for cross coupling. As an example, consider traces over a ground plane as in Figure 1.2. The ground plane can be considered the reference conductor at zero volts.

Assume that one of the traces is at a voltage V_1 and all the other conductors are at zero volts. By definition the self-capacitance C_{11} of trace 1 is the ratio of charge on trace 1 to the voltage on trace 1 or Q_1/V_1. When trace 1 is at potential V_1, some of the electric field lines terminate on trace 2. This represents an induced charge Q_2 on this conductor. The ratio of $Q_2/V_1 = C_{21}$ is called a *mutual capacitance*. Since the field lines always terminate on charges of opposite polarity this ratio is always negative.

N.B.

Mutual capacitance is often called a *leakage* or *parasitic capacitance*.

Mutual capacitance can allow an induced current to flow in nearby circuits. The coupled current flow depends on the rate of change of voltage and the resulting voltage depends on the victim circuit impedance. This coupling can be a problem when traces

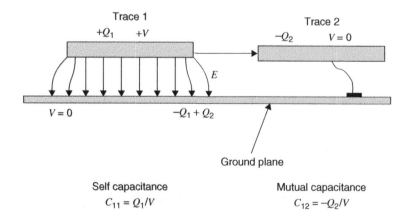

Figure 1.2 Mutual capacitance between traces on a circuit board.

on a circuit board run in parallel for a long distance. Mutual capacitance can be measured by placing a sinusoidal voltage on one conductor and observing the current flow on a second conductor. A small series resistor can be used to sense the current in the second conductor.

N.B.

When a self-capacitance measurement is made on a circuit board, all other conductors must be grounded.

Later, we will show that in cross coupling, coupling current can be introduced only when a logic signal is in transition.

1.9   *E* FIELDS INSIDE CONDUCTORS

If there is current flow in a conductor, a small *E* field must exist inside that conductor. At dc, this field will be the same at any depth. In most of our applications the ratio of external to internal field strength is about 20,000:1. A changing magnetic field associated with current flow forces the current to stay near the surface of the conductors. This applies to logic operations, where currents flow very near to the surface of the conductors. This is known as *skin effect*. For a sine wave current at 1 MHz, the skin depth for copper is only 0.066 mm. This means that the current at this depth is the surface current reduced by 8.68 dB. At 100 MHz, the skin depth is 0.0066 mm. Most of the current flows in the first three skin depths.[2]

N.B.

Very small *E* fields are required to cause typical current flow in conductors.

When a step voltage is first applied to a trace over a ground plane, the charges that move are on the surface of that trace and on the surface of the ground plane under the trace. Consider a fixed point on a long transmission line. As time progresses, the current that flows begins to penetrate into both conductors. After a few milliseconds, there will be very little skin effect present.

---

[2]Published skin depth equations usually relate to plane electromagnetic sine waves impinging on an infinite conducting plane. The depth that is calculated by this method is generally used as an approximation for other conductor geometries. As an example, it provides a rough measure of the current penetration in a round conductor, a trace or in a conducting plane.

N.B.

Current patterns in traces and ground planes are very complex. Fortunately, in digital circuits, these patterns do not require extensive analysis. Current only flows in a conductor where there are fields present.

1.10 THE D FIELD

A second description of the electric field is called the D or *displacement field*. This field starts and stops on charges. Unlike the E field, the D field does not change intensity at the boundary of a dielectric unless there are surface charges present.

A changing D field is equivalent to a current flowing in space. This current is usually called a *displacement current*. In a capacitor, the D field between the plates of the capacitor changes as the charge on the surface of the plates changes. By using this interpretation, the changing D field is an extension of the current that flows in the connecting conductors. In circuit theory, current flows in a capacitor even though the plates of the capacitors are insulated from each other. In antenna theory, the current that flows in the antenna structure returns as displacement current in the nearby space.

The relationship between the D and E field is given in Equation 1.4 as

$$D = \varepsilon_0 \varepsilon_R \cdot E \tag{1.4}$$

where ε_0 is the permittivity of free space or 8.85×10^{-12} F/m and ε_R is the relative dielectric constant.

1.11 ENERGY STORAGE IN A CAPACITOR

The work required to move a unit of charge q across the plates of a capacitor is equal to qV, where V is the voltage across the plates of the capacitor. When the capacitor is discharged there is no voltage and the work required to move the first unit of charge is zero. When the second unit of charge is moved it must do work against the field created by the first unit of charge. The work required to move the last unit of charge is qV. On average, the work done in moving a total charge Q is $1/2QV$.

The energy stored in a capacitor can be expressed in terms of the capacitance by noting that the ratio of Q/V is the capacitance C. Using this relationship, the energy in a capacitor is given by

$$E = \frac{1}{2}CV^2 \tag{1.5}$$

1.12 THE ENERGY STORED IN AN ELECTRIC FIELD

The capacitance C in Equation 1.5 is determined by the geometry of the conductors. If we use Equation 1.1 and the definition that the E field is V/d, it is easy to show that the energy stored in the capacitor is

$$E = 1/2\,E^2\,\frac{V}{\varepsilon} \tag{1.6}$$

where V is the volume of space containing the E field and ε is the dielectric constant.

Equation 1.6 is quite general and applies to an E field in a capacitor, in free space, or associated with a transmission line. In most cases the E field varies with position. An integration is needed to calculate the total E field energy in a volume of space.

1.13 THE MAGNETIC FIELD

The magnetic field that exists around a current is called an *H field*. This field is a force field that reacts against another magnetic field. The force field can be observed using a small compass or by noting the alignment of iron filings on a piece of paper that is threaded by a current carrying conductor. The Biot Savart's (Ampere's) law says that the line integral of H around any closed path is $4\pi I$, where I is the current threading that loop. The value of H along the path of integration depends on permeability. In a high permeability part of the path, H is a low intensity field.

The H field is a vector field as it has intensity and direction at all points in space. The shape of the H field is usually represented by a series of curved lines. The H field intensity is the greatest where these curved lines concentrate. For a single round conductor carrying a steady current, the H field is represented by circles that surround the conductor (see Fig. 1.3 in Section 1.18).

The second description of the magnetic field is the induction or B field. The induction measure of the field is needed whenever voltages are involved. The value of B in teslas is $\mu_0\mu_R H$, where μ_R is the relative permeability of the material and μ_0 is the permeability of free space equal to $4\pi\,10^{-7}$ teslas-meters per ampere. The voltage induced in any conducting loop of area A is the rate of change of the induction flux or $A\,dB/dt$. This is known as *Faraday's law*.

The flux generated by the induced current is always opposite in direction to the flux inducing the current. This is known as *Lenz's law*. The B field intensity does not change value at the boundary between regions of different permeability. The B field is always represented by closed continuous curves. At a boundary where the permeability changes, the B field may change direction but not intensity.

In electrostatics, a conductor geometry that is designed to store an E field is called a *capacitor*. In magnetics, a conductor geometry that is designed to store a B field is called an *inductor*. The definition of inductance is given as the B field flux generated per unit of current. The magnetic flux ϕ in any loop is the product of B field intensity and the area perpendicular to the direction of the flux.

The energy E stored in a magnetic field comes from accelerating charges into a current path creating the field. Consider an inductor with no current flow in the coil. At the start there is no magnetic field so no work is required to move the first increment of current into the coil of the inductor. For this analysis we must assume the resistance of the coil is zero ohms. When the second increment of current is added there is an existing magnetic field. The work required to establish the field in a volume V, where the field intensity is B, is

$$E = 1/2 B^2 \frac{V}{\mu} \tag{1.7}$$

In any practical geometry the B field is not constant, and the total energy must be calculated by an integration.

A magnetic field can be associated with an inductor or in the space around any current path. A magnetic field can be located in free space as a part of the radiation or it can be a part of the field in motion on a transmission line.

An inductor is a conductor geometry designed to store magnetic field energy. Measuring magnetic flux directly is very difficult, and, for this reason, inductance is usually measured by noting the voltage in the relation

$$V = L \frac{dI}{dt} \tag{1.8}$$

where dI/dt is the rate of change of current with respect to time. If V is in volts and dI/dt is in amperes per second then the inductance has units of henries. It is important to consider inductance as a field concept where the flux created by current flow is in the space around conductors. This is where the field energy is stored.

N.B.

Inductance implies the ability to store magnetic field energy. Conductor geometries designed to store this energy are called *inductors*.

In a transmission line, the flow of current implies an associated magnetic field. The inductance per unit length of line will be an important factor in describing transmission lines. When a transmission line carries energy, there must be both an electric and a magnetic field in the space between the conductors.

N.B.

The energy that is carried by any pair of conductors is carried in the space between the conductors. Both the E and H fields must be present, if energy is in motion.

1.14   RISE TIME/FALL TIME

The typical logic signals we will encounter are voltages that step between two values. The one value is approximately the power supply voltage V and the second value is approximately the zero reference. For most of the discussions in this book we use the power supply voltage and zero volts. Logic is usually controlled by a clock signal that is a square wave. The rise and fall time of this clock signal is usually less than 10% of the clock cycle. A 100 MHz clock rate might have rise and fall times of 1 ns. The rise and fall time of the logic signals should be in this same range.

The Fourier spectrum of a square wave consists of sine waves at the fundamental frequency and at all odd harmonics of that frequency. The amplitude of the fundamental harmonic is the voltage $V_1 = V/\pi$ volts rms. The amplitude of the nth harmonic is $V/\pi n$ volts rms. A 10-MHz square wave is made up of sine waves at 10, 30, 50, 70, ... MHz. The amplitudes of these sine waves are $V/3.14, V/9.42, V/15.75, V/21.98, ...$ When these sine wave voltages are added together, a square wave of voltage V is the result.

In a practical circuit, a square wave has a finite rise or fall time. The harmonics (sine waves) that make up this square wave will have amplitudes that fall off proportional to frequency out to a frequency of $1/\pi \tau_r$, where τ_r is the rise time or fall time, whichever is the smallest. Beyond this frequency, the harmonic amplitudes fall off proportional to the square of frequency. If the 10-MHz square waves have a rise time of 10 ns then the harmonics above 30 MHz are not a serious problem.

It is impractical to analyze logic signals using sine waves. Logic signals are not perfect square waves and the length of the transmission lines is an important factor that is often not considered. The rise time (fall time), however, does provide key information that will be very useful in later discussions. It is important to understand the ideas of harmonic analysis as this is often the only tool we have to solve certain types of problems.

1.15   MOVING ENERGY INTO COMPONENTS

To place energy into a capacitor, charges must move. These moving charges are current, and this current implies a magnetic field. To move current into an inductor, a voltage is required. This means both a magnetic and an electric field are needed to move field energy into a capacitor or an inductor. This implies that all electrical activity requires the presence of both an electric and a magnetic field. Analog circuits (below 100 kHz) usually require small currents that change level slowly. For this reason, the magnetic fields associated with low frequency analog signal transport can usually be ignored. For the material in this book where clock and logic signals are changing rapidly, the magnetic field plays a very important role.

All passive and active electronic components require fields to operate. The problem for the circuit designer is to bring these fields to the components over the interconnecting conductors. As we have just stated, these fields carry energy. At low frequencies, there are few problems of field transport. For most analog circuits, the current levels are in the milliamperes. This is not true for digital circuits where amperes may be involved. These currents may need to flow for only a few nanoseconds, but if the energy is not available, the circuit will not function.

The problem we will discuss at great length is supplying the energy that must be carried through logic switches and on traces to other parts of the circuit. This energy can be stored on local capacitors, between power and ground planes and often under logic traces.

1.16   FARADAY'S LAW

When the changing flux of a magnetic field crosses a conducting loop, there is a voltage induced in that loop. The voltage depends on the rate of change of that flux. In digital circuits, the wave energy moving down a transmission line involves a changing current. This changing current results in a changing H field. To calculate the induced voltage, the B field measure must be used. If the induction B field is in units of henries, then the coupled flux is BA, where A is loop area coupled to the flux. The induced voltage is therefore

$$V = \mu A \frac{dH}{dt} \tag{1.9}$$

where A is the coupling area in meters squared. The degree of coupling depends on circuit geometry and how rapidly the fields are changing. We will discuss parallel trace spacing and trace routing, as it relates to cross coupling, in later chapters.

1.17   SELF- AND MUTUAL INDUCTANCE

The terminology used in describing mutual inductance is very similar to that used to describe mutual capacitance. When magnetic flux is generated in a circuit, some of the flux crosses into other circuits. For an ideal inductor, the flux that is generated stays in that component. The ratio of flux to current in one geometry is called *self-inductance*. This is the fundamental definition of inductance.

The definition of mutual inductance is the magnetic flux coupled into a second circuit for a current flowing in a first circuit. The coupling can be of either polarity and depends on conductor geometry. Our interests will be the coupling between traces on circuit boards. The notation that is used is the double subscript such as L_{11}, which means the ratio of magnetic flux generated in loop 1 by current in loop 1. The notation L_{12} means the ratio of flux in loop 2 to the current in loop 1.

N.B.

The inductance per unit length of a circuit trace can be derived from the capacitance per unit length by recognizing that the velocity of transmission is $(LC)^{-1/2} = c/\varepsilon_R$, where c is the velocity of light.

The cross coupling that occurs in digital circuits goes in both directions. This coupling involves both electric and magnetic fields. This problem is discussed in Chapter 3.

1.18 POYNTING'S VECTOR

When electrical energy moves, both an E field and an H field are required. The power density at a point in space is given by the vector cross product of the E- and H field intensities. This vector P in Equation 1.10 is called *Poynting's vector*. This vector exists at

$$P = E \times H \qquad (1.10)$$

all points in space and points in the direction of energy flow. For a pair of conductors carrying power from a battery to a load, the E field crosses between the conductors and the H field circles the current carrying conductors (Fig. 1.3).

Note that the E and H fields have directions perpendicular to the current path and perpendicular to the direction of power flow. The E field has units of volts per meter and the H field has units of amperes per meter, so the product has units of watts per meter squared.

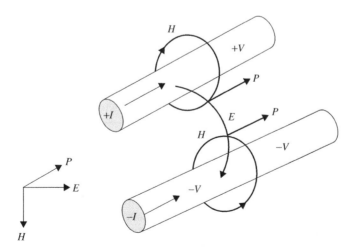

Figure 1.3 Poynting's vector for a pair of conductors carrying energy. E and H field vectors exist at all points in space. $P = E \times H$. Note that P shows the density and direction of power flow at every point in space.

N.B.

The energy supplied to a load is carried in the space between the connecting conductors and not in the conductors. This concept applies at all frequencies including dc. The P vector is greatest at points near the conductors.

It is important to view every pair of conductors as a transmission line. Fields carry the energy in the space between conductors at all frequencies. Fields that leave the circuit represent radiated energy. In this case, a component of Poynting's vector points in the direction of radiation. When there is heat loss, then a component of Poynting's vector points to the region of dissipation. For a round conductor, this vector component points to the center of the conductor.

Note the following:

- A transmission line can carry energy in both directions.
- A transmission line can transport any number of signals at the same time.
- A transmission line is a path that allows energy to flow to a lower energy state.
- There is no way to tell if the field is carrying signal or energy. They are the same.

The transmission lines we will consider are the traces on circuit boards that make connections to capacitors, vias, conducting planes and cables and that connect logic signals between logic elements.

N.B.

A building requires walls and halls. People move in the halls not in the walls. A circuit board has traces and spaces. Energy moves in the spaces not in the traces. The halls direct where people can go. The traces direct where the energy can go.

1.19 FIELDS AT DC

The fields associated with traces over a conducting plane are located under the traces. At frequencies below a kilohertz, a circuit analysis will show that the series inductive reactance between points is lower than the resistance between points. At these low frequencies, current patterns are controlled by conductor resistances. The field required to move current in a ground plane at dc is very small. As a result, the E field pattern at dc is very similar to the pattern at 1 MHz. If the resistivity of the ground plane were zero then no field would be required to move current and the pattern under the trace would not depend on frequency.

GLOSSARY

B **field** (Section 1.13): The B field is the field of magnetic induction. The unit of B field intensity at a point is the tesla. The B field is a vector field, as it has intensity and direction at every point in space. The B field does not change intensity across a boundary when the permeability changes value. When the B field is represented by lines, these lines always form closed loops. The B and H fields are related by permeability $B = \mu_0 \mu_R \cdot H$, where μ_0 is the permeability of free space or $4\pi 10^{-7}$ and μ_R is the relative permeability.

Capacitance (Section 1.7): The ability of a physical space to store electric field energy. In circuit applications it is the ability of a conductor geometry to store electric field energy (capacitor). If the electric field energy is not contained by conductors, it must be moving at the speed of light. Examples of fields in motion are transmission lines or antennas. Space has a capacitance per unit volume.

Capacitor (Section 1.7): A circuit component designed to store electric field energy.

Charge (Section 1.4): A quantity of electrons (negative charge). The absence of electrons (positive charge). The unit of charge is the coulomb. A coulomb of charge flowing past a point in one second is an ampere.

Charged (Section 1.4): A mass that has an excess or a lack of electrons on its surface.

Current (Section 1.6): The flow of electrons or ions. Electrons can flow in a conductor. Ions can travel in space or in a chemical solution.

D **field** (Section 1.10): The electric field representation that originates and stops on charges and is independent of the dielectric constant. The D field is continuous between charges located on the surface of separate conductors. The D field and the electric field are related by the dielectric constant $D = \varepsilon_0 \varepsilon_R E$, where ε_0 is the permittivity of free space and ε_R is the relative permeability. See the definition of permittivity.

Displacement current (Section 1.10): When an electric field changes intensity, it is equivalent to a displacement current in space. It is not electron flow. As an example, this current flows in the space between the plates of a capacitor when the voltage across the capacitor changes. A displacement current has an associated magnetic field.

E **field** (Section 1.5): The force field that exists around all electric charges. When the charges are located on physical masses, the forces appear to act on the masses. Charges that create an electric field may be located in space, on the surface of conductors, or trapped inside an insulator (dielectric). An E field can exist without the presence of fixed charges when it is moving at the speed of light. As an example, the E field that leaves an antenna or flows down a transmission line. The E field intensity at a point in space is measured in units of volts per meter. The E field has intensity and direction at every point in space. It is a vector field.

Fall time (Section 1.14): The length of time it takes for a signal to go from 80% to 20 % of its initial value. Fall time applies to step changes in voltage, current, or field intensity.

Faraday's law (Section 1.16): A voltage is induced in any loop by the changing magnetic flux crossing that loop.

Field energy (Section 1.13): For a region where the E field is constant, the energy stored is $1/2\ E^2 V/\varepsilon_0$, where V is the volume of space. In a region where the H field is constant, the energy stored is to $1/2 H^2 V/\mu_0$, where V is the volume of space. See Poynting's vector.

Fourier spectrum (Section 1.14): The harmonic content that makes up any repetitive wave form. A single event has a spectrum that exists at all frequencies. There is no energy at any one frequency. For a single event, it is practical to discuss the voltage in a band of frequencies.

Ground (Section 1.5): A conducting surface that is large compared to the components or conductors located nearby. A conducting surface designated as the reference or zero of potential.

H field (Section 1.13): The magnetic field around a element current. The force of a magnetic field is against a second magnetic field. When two conductors carry current, the forces are felt on the conductors. If the H field is not confined by conductors, it moves at the speed of light. Examples are transmission lines and antennas. The H field has units of amperes per meter. It is a vector field as it has intensity and direction at every point in space. The H field is discontinuous at transitions in permeability. Around a current carrying conductor, the value of H is given by Ampere's law as $I/2\pi r$, where r is the distance from the conductor.

Inductance (Section 1.13): The ability of a physical space to store magnetic field energy. Inductance has units of henries. The inductance of a conductor geometry is the ratio of total magnetic flux generated per unit current. In circuit applications, it is the ability of a conductor geometry to store magnetic field energy. If this field energy is not sustained by conductors, it must be moving at the speed of light. Examples are transmission lines and antennas. Space has an inductance per unit volume.

Inductor (Section 1.13): A conductor geometry designed to store magnetic field energy for the flow of current.

Lenz's law (Section 1.13): The flux generated by an induced current is always opposite in direction to the flux generating the current.

Magnetic field energy (in an inductor) (Section 1.13): Field energy can be introduced into an inductor by accelerating charges. It takes work to move a charge into an existing current path that creates the magnetic field. The total work is $1/2 L I^2$.

Mutual capacitance (Section 1.8): The ratio of charge induced on a second conductor by a voltage on a first conductor when all other conductors are at zero potential.

Mutual inductance (Section 1.17): The magnetic flux coupled to a second conductor loop from a current flowing in a first conductor loop. The resulting loop voltage can be of either polarity.

Permeability of free space: The ratio between the B and H fields in free space. If B is in tesla and H is in amperes per meter then $B = \mu_0 H$, where $\mu_0 = 4\pi \times 10^{-7}$.

Permittivity of free space (Section 1.10): The ratio between the D field and E field in free space. $D/E = \varepsilon_0 = 8.854 \times 10^{-12}$ F/m.

Poynting's vector (Section 1.18): The vector cross product of E and H at a point in space. It represents the power crossing per unit of area and the vector points in the direction of power flow at this point in space.

Rise time (Section 1.14): The length of time a signal takes to go from 20% to 80% of its final value. Applies usually to step functions of voltage.

Skin effect (Section 1.8): The flow of current on the surface of conductors at high frequencies. The magnetic field from a changing current keeps that very current from entering the conductor.

Self-capacitance (Section 1.7): The ratio of charge on a conductor to voltage on that same conductor with all other conductors at zero volts. The capacitance C of a capacitor.

Self-inductance (Section 1.13): The ratio of magnetic flux generated by a circuit to current in that same circuit. Usually given the symbol L. The voltage across an inductor is $L \, dI/dt$. This is usually the definition of inductance as flux cannot be measured directly.

Tesla (Section 1.13): The unit of magnetic field intensity for the B field.

Transmission line (Section 1.1): Any two parallel conductors, one of which can be a conducting plane.

Voltage (Section 1.5): (Properly a voltage is a potential difference) The work required to move a unit electrical charge between two points. In most circuit applications, these end points are on conductors and the work is done in the space between the conductors. Work is the product of force and distance when the force is in the direction of motion. The force that does work on the unit electrical charge can only result from the two electric fields. A voltage difference can exist between any two points in space and between points on conductors.

2

TRANSMISSION LINES

2.1 INTRODUCTION

The language of transmission lines was developed in the early days of radio when rf (radio frequency) signals had to be transported from a transmitter to an antenna. The signals were basically a modulated sine wave called a *carrier*. The problem was to match impedances so that the transported energy reached the antenna and was effectively radiated. In digital electronics, the signals are step functions and the hope is that very little of the energy is radiated. Signals are often transported over and between conducting planes, and this practice was not a prime consideration in early radio.

Transmission line theory uses many terms common to circuit theory. We will be working mainly with step functions and not sine waves. A transmission line can be considered a distributed parameter circuit. Words such as bandwidth and frequency spectrum will apply. We will often use the term *impedance* but not in a sinusoidal sense.

Digital Circuit Boards: Mach 1 GHz, First Edition. Ralph Morrison.
© 2012 John Wiley & Sons, Inc. Published 2012 by John Wiley & Sons, Inc.

> **N.B.**
>
> Circuit traces on a printed circuit board are transmission lines; and circuit traces over or between conducting planes are transmission lines.

When a voltage is placed on a transmission line, it is an invitation for nature to use this conductor geometry to move available energy away from the source. At the moment of connection, both voltage and current begin their travel down the line. We will often call this voltage and current flow a *signal* or a *wave*.

When a signal (wave) propagates down a transmission line, the source of signal has no way of sensing the length of the line or the type of termination. Energy is sent down the line just as if it was going to be radiated by a terminating antenna. In fact, a wave must reach a discontinuity, reflect and return to the source before the source can react. This type of delayed reaction is not usually considered in linear circuit theory.

The energy that is transported on a transmission line requires the presence of both an electric and a magnetic field. It is correct to consider a transmission line

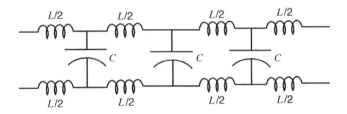

Figure 2.1 The lumped parameter model of a transmission line. *L*, inductance per unit length; *C*, capacitance per unit length.

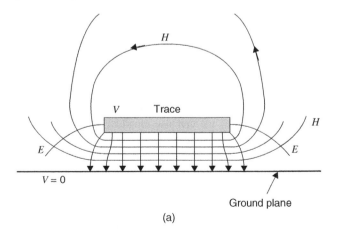

Figure 2.2 (a) The field pattern around a transmission line.

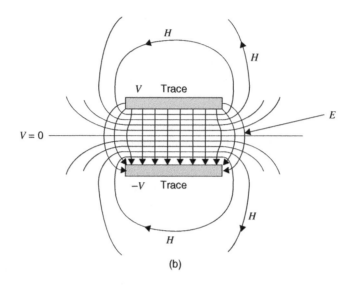

Figure 2.2 (b) The field pattern around a two-trace transmission line.

as a distributed inductance and capacitance. A schematic representation that shows lumped parameters would imply that the energy in the line is stored inside these pseudo components. In practice, the field patterns are around the conductors and extend out into space. This is the reason there is some radiation. The lumped parameter model is shown in Figure 2.1 and the field pattern is shown in Figures 2.2(a) and (b). E and H fields exist in all the space around the traces. In regions where the intensity is low, lines may not show.

2.2   SOME COMMON ASSUMPTIONS

The voltage symbol that we will use in this book implies an unlimited source of current with zero rise time. This perfect voltage source does not exist, but it will be used to avoid adding complications. An instantaneous voltage implies an instant electric field. Since this field stores energy, the generation of a field in zero time requires infinite power. Similarly, the instantaneous flow of current implies a magnetic field and again this field stores energy. The generation of field energy in zero time would again imply infinite power. If we are careful, we can use an ideal step voltage source or a step current, and we will not get into trouble.

N.B.

Connections to a power supply require some length of wiring. This wiring is a transmission line. For this reason alone, there are no sources of voltage with zero source impedance.

> **N.B.**
>
> Every capacitor has lead length associated with its internal conductor geometry. The internal voltage of the capacitor is always in series with some form of transmission line.

The ideal switch does not exist. Even if the switch were mechanical, the capacitance before the moment of contact must get very large. If the switch is solid state, the closure is really a nonlinear dynamic impedance. We will use the ideal switch in our discussions even though it does not exist.[1]

A resistor at the end of a transmission line is called a *termination*. A resistor termination is a complex network because every resistor has a distributed series inductance, a parallel distributed capacitance, and skin effect. We use these circuit terms to make the point that every component is complex if examined closely. In microwave work, a wave guide termination is not a simple resistor. A valid terminating resistor must be planar in character and it must cover the entire end of the guide. We will often be terminating a transmission line with a resistor where picoseconds are of interest. One picosecond is one cycle at 1000 GHz, which is well above microwave frequencies. When we show a resistor on a schematic, we will assume it is a valid resistor at the frequencies of interest.[2] It is important to keep in mind that we are always making assumptions. Ideal components do not exist except in our equations and maybe only in our minds. We need these ideal elements to simplify our analysis. We need these ideal elements to start any sort of discussion.

2.3   TRANSMISSION LINE TYPES

The transmission lines on a typical circuit board are traces over a ground or power plane or traces placed between conducting planes. Parallel traces can also be a transmission line carrying a balanced signal (odd-mode transmission). This transmission line arrangement allows circuitry to reject interference, which is common-mode in character (even-mode transmissions). Parallel traces can be isolated, over a ground plane or in a conducting sheath.

---

[1]IC switches (drivers and receivers) are modeled by IBIS (I/O Buffer Information Specification). This information is controlled by ANSI/EIA-656 (American National Standards Institute). Information is supplied by IBIS because manufacturers often want to withhold proprietary circuit information.

[2]Small carbon resistors are used in practical logic circuit design for impedance matching. If the parallel parasitic capacitance is 2 pF and the R is 50 ohm, the RC time constant is 1 ns. In gigahertz designs, 1 ns is one clock cycle. In series terminations, the effect is to enhance the leading edge of a switched signal. This effect is often ignored in any calculation. A resistor is also a short section of poorly designed lossy transmission line. Including this complication in any analysis would make our approach unnecessarily difficult. For this reason, we assume for now that a resistor is not another transmission line.

A coaxial cable can be a transmission line. The conducting sheath is the return line. In some cases, two carefully positioned conductors in a shielded enclosure can serve as a balanced transmission line. A balanced structure means that any interference couples to each line equally. It is possible to reject this common signal in the receiving electronics.

Traces between conducting planes are also transmission lines with fields that are confined by the planes. These planes reduce radiation and limit coupling from external fields. Any pair of conductors that form a parallel path for signals can be considered a transmission line, although the pair may not be practical for gigahertz transmissions.

All conductor pairs are capable of transporting electromagnetic energy in both directions. Pairs include conductors in a cable, shield-to-shield, cables-to-conduit, cables over rack surfaces, power lines over the earth, etc. In this book, we direct our efforts toward traces and conducting planes on circuit boards. Transmission lines that connect to a board from external sources can transfer interference into and out of the board. Interference can be rejected by using balanced transmission techniques and, in some cases, passive filters. Interfacing cables with circuit boards is discussed in later chapters.

The energy that is transported on any transmission line is in the field between the conductors. When one of the conductors is a conducting plane, the fields associated with the energy are confined to the area immediately under that trace.[3] The current that flows is confined to the trace and the conducting surfaces directly under the trace. The current path follows the surface areas where the field patterns terminate. The current path in the plane can be likened to a river under the shadow of the trace. Even though one conducting plane may be shared by many traces, there is no cross coupling, unless the fields associated with transmission share the same volume in space. A simple right-angle crossing of traces is acceptable. The rivers of current (fields) will know what to do.

N.B.

In circuits, most of the field energy is stored in space not in conductors.

Transmission lines formed by traces over a conducting plane are called *microstrips*. If a dielectric surrounds the trace, the term *embedded microstrip* is used. This type of transmission line is usually found on the outer layers of a PCB (printed circuit board). Sometimes the embedding dielectric can be a conformal coating or a

---

[3]The field pattern associated with a fixed logic signal on a trace over a ground plane is nearly constant with time. There are changes to the field pattern associated with skin effect. As the wave penetrates the conductor, the current pattern changes slightly. The field energy inside the ground plane or in the trace is very small. When there is a step demand for energy, it comes mainly from the E field in the space under the trace.

screening material. See later discussions on characteristic impedance. Transmission lines formed by traces between two conducting planes are called *stripline*.

N.B.

For digital logic, the return current for a wave in progress flows on the conducting plane directly under the logic trace. The return current does *not* represent the return of energy. This current direction is necessary to support the direction of the magnetic field. When there is a reflection, the second wave is superposed on the (field) voltage of the initial wave. The current of the reflected wave flows in the same two conductors.

N.B.

The field pattern between two traces spaced at distance d apart is double the field pattern of one trace spaced at distance $d/2$ over a conducting plane.

N.B.

Logic currents that flow on a conducting plane stay on the surface of that plane under the trace. Currents (fields) cannot cross to the other side of the plane except though a hole (via) or at the board edges.

2.4 CHARACTERISTIC IMPEDANCE

Figure 2.3 shows how a wave propagates when a voltage with a finite rise time is switched on to a length of transmission line.

This line can be characterized as having an inductance and a capacitance per unit length. The step wave that propagates down the line will place charges on the distributed capacitance at a fixed rate. After the wave is established, a steady current is supplied by the voltage source. A steady voltage and a steady current imply that the transmission line looks like a resistance. This resistance value is called the *characteristic impedance* of the line. If the line is infinite in length, the current–voltage relationship will be given by Ohm's law as

$$Z_0 = \frac{V}{I} \tag{2.1}$$

where Z_0 is the characteristic impedance.

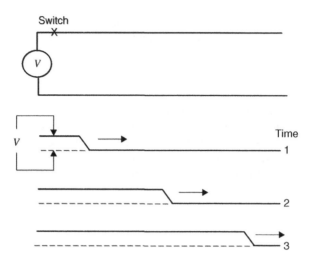

Figure 2.3 How a step voltage propagates on an ideal transmission line.

In theory, the voltage V can have any wave form. In digital circuits, logic voltages are step functions with finite rise and fall times. In rf (analog) circuits, the signals are usually sine waves.

The characteristic impedance of a transmission line is given by

$$Z_0 = \left(\frac{L}{C}\right)^{1/2} \tag{2.2}$$

where L and C are the inductance and capacitance per unit length, respectively. In circuit board designs, the characteristic impedance of each logic trace must be considered. This practice provides uniformity in terms of reflections, current demands, and losses. Very accurate control is not practical, as there are many side effects that influence the impedance. For example, the dielectric constant will vary between boards and in different areas of the same board. Nearby traces will modify the impedance. Trace dimensions will vary during manufacturing, and this directly influences the characteristic impedance. The amount of dielectric material around a trace will vary with location. Dielectrics can include conformal coatings, coupons, and, to some extent, moisture.

The characteristic impedance of a transmission line is equal to the ratio between the E and H fields at every point in the space near the conductors. This ratio holds for every wave that is in motion on the line. The ratio of E to H inside of traces is different. Stating this in another way, a transmission line will only support waves with a fixed wave impedance.

Equations for the characteristic impedance of trace geometries are available from many sources. These sources include trade publications, books, the internet, engineering service organizations, and hardware planners. A few equations are supplied in Chapter 6 of this book. Typically, traces on a circuit board have characteristic impedances around 50 ohm. This value turns out to be effective for most designs. Higher impedances invite cross talk and lower impedances place extra demands on the current supplied by the logic. For some logic families, the optimum characteristic impedance may be other than 50 ohm. It is perfectly acceptable to design a board where different trace groups have different characteristic impedances.

2.5 WAVE VELOCITY

The velocity of a wave traveling on a transmission line is given by

$$v = (LC)^{-1/2} \qquad (2.3)$$

where L and C refer to the inductance and capacitance per unit length of line, respectively.

This is equal to the velocity of light divided by the square root of the relative dielectric constant or

$$v = \frac{c}{\sqrt{\varepsilon_R}} \qquad (2.4)$$

N.B.

The wave velocity is usually measured at the leading edge of the wave. This is somewhat inexact as the rise time tends to increase with transit time.

For circuit boards where the relative dielectric constant is ~4.0, the wave velocity is 0.015 cm/ps. This means a wave travels in a board dielectric about 1/16 inch in 10 ps.

The dielectric constant and the losses along a transmission line will vary as a function of frequency. This means that the sine wave voltages that make up the initial wave do not travel at the same velocity. As the wave travels, the wave front will lose its sharp character. The amount of wave front distortion will depend on the type of dielectric, on line losses, on radiation, and on line length. Board material that is rated to operate at a higher frequency will distort the wave less over a given length of trace.

In microstrip transmission lines (outer traces), a part of the E field is external to the dielectric, which means that some of the field travels at a higher velocity. This distorts the leading edge for short rise time transmissions. This has the effect of increasing cross talk between traces on an outer layer. Traces that are buried in a dielectric or that are routed between conducting planes (stripline) do not have as large a difference in wave velocity (see the discussion on cross talk in Chapter 3).

If the characteristic impedance of a transmission line is known, the inductance per unit length can be determined by dividing Equation 2.2 by Equation 2.3 or Equation 2.4.

The capacitance per unit length can be determined by multiplying Equation 2.2 by Equation 2.3 or Equation 2.4.

2.6   STEP WAVES ON A PROPERLY TERMINATED LINE

When an ideal transmission line is infinite in length there are no reflections. If the line is finite but terminated in a resistance equal to its characteristic impedance, the line will behave as if it is infinite in length. The ideal step wave action along this line is shown in Figure 2.4. After the switch closes, a wave (1) progresses down the line.

This resistor placement is usually called a *parallel termination*. The terminology *shunt termination* is also used (see Section 2.17 for series terminated lines).

When the step wave (1) reaches the end of the line, the voltage and current match the requirements of the load, and all wave action stops. There is no radiation after the wave reaches the termination. This wave process occurs whether the transmission line is a trace over a conducting plane or simply two parallel traces. Energy flows in the field between the conductors and not in the conductors themselves. The traces are guides. The analogy with train tracks is appropriate. A train can only go where there are tracks. The only things that leave the track are passengers, heat, and smoke. Then there is the train wreck, where everything leaves the track.

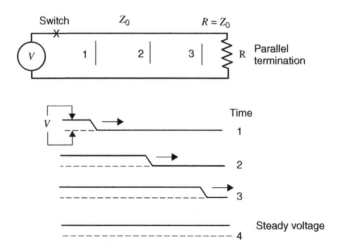

Figure 2.4   The wave associated with a line terminated in its characteristic impedance.

2.7   THE OPEN CIRCUITED TRANSMISSION LINE

Consider the open circuit transmission line in Figure 2.5a. The diagram is a one-line representation of a two conductor circuit as shown in Figure 2.4. When the switch closes, a step wave (1) of voltage V progresses down the line. When wave (1) reaches the end of the open-ended transmission line, a reflected wave (2) must cancel the current at the end of the line. Poynting's vector for wave (2) requires that only the H field in this wave must reverse in direction. As wave (2) progresses back on the line,

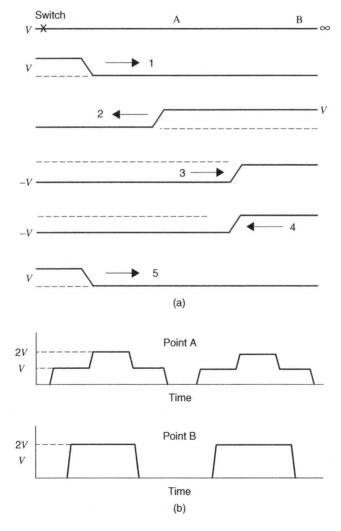

Figure 2.5   (a) The sequence of individual waves generated on an open line after a switch closure and (b) the wave forms at points A and B.

the currents sum to zero. The voltage of wave (2) has the same polarity and voltage as the initial wave. Wave (2) simply adds to the voltage placed on the line by wave (1). The result is that the voltage on the line doubles, as wave (2) returns to the voltage source. It is important to note that energy is flowing from the source from the time the switch closes until wave (2) returns to the source.

When wave (2) reaches the source, the voltage is incorrect. The reflection of wave (2) is wave (3) that moves forward on the line. It is a negative voltage $-V$. The sum of waves (1), (2), and (3) is simply V. The current for wave (3) is supplied by wave (2). This means that zero current is supplied by the source voltage. From this point on, the source does not supply any more energy to the line. The total energy supplied to the line is $2V^2 \, t/Z_0$, where t is the transit time from the source to the open circuit.

When wave (3) reaches the open end of the line, it is again reflected. Since wave (3) is a negative voltage, the reflection must be of the same polarity. The result is that the voltage at the open end of the line is again zero. This assumes no losses on the line.

The voltage at any point on the line is the sum of all the wave voltages that have gone past this point at an earlier time. The voltage at the open end of the line at B is a step wave of double amplitude. The wave form at the midpoint A along the line is a voltage in the sequence $0V, V, 2V, V, 0V, V, \ldots$ These voltage wave forms are shown in Figure 2.5b.

N.B.

The voltage at the source does not change value. The voltage at the open end of the line is either doubled or zero. On a typical trace, the conformal coating must be removed if the voltage waveforms are to be observed along the line.

On an ideal line without losses, this back and forth wave action will continue indefinitely. The energy stored on the entire line is constant except for line losses.

The load on the end of a transmission line may be a logic gate. In many of our discussions, we will treat a logic gate as an open circuit. A gate connection usually looks like a small capacitance, which has the effect of slowing the rise or fall time. A slower rise time reduces both radiation and losses in the dielectric. For a small terminating capacitance, the voltage at the open end of the line will still double, and there can still be radiation along the line. The conductors and the logic switch will dissipate energy as heat as long as there is current flow. There will be radiation along the line where voltage is in transition. In practice, each time the wave makes a round trip; the leading edge will be further attenuated and distorted.

If the terminating capacitance is large then the lumped inductance of the transmission line will resonate with the capacitor resulting in significant ringing. This is circuit behavior.

N.B.

A capacitive termination does not necessarily eliminate voltage doubling. It simply slows the doubling process. On short transmission lines where the rise time is long, voltage doubling may not take place (see Section 3.7 for the rule governing voltage doubling).

N.B.

There is no way for the field energy placed on a transmission line to return to the driving circuit or to the power supply. It must eventually be dissipated in heat or it must radiate.

2.8 THE SHORT CIRCUITED TRANSMISSION LINE

Consider the short circuited transmission line shown in Figures 2.6(a) and (b).

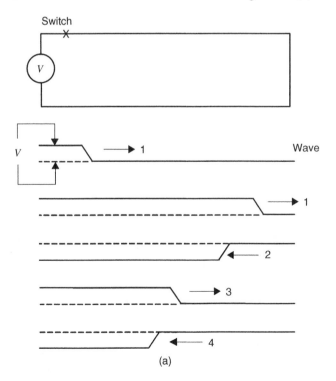

Figure 2.6 (a) Voltage on a transmission line with a short circuit termination.

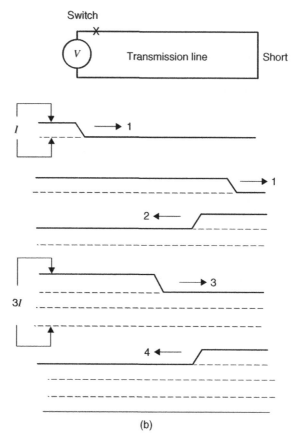

Figure 2.6 (b) Current waves in a transmission line with a short circuit termination.

After the switch closes, a step wave (1) travels down the line to the right. At the short circuit, a reflected wave (2) starts back that cancels the voltage on the line. Poynting's vector on wave (2) requires that the E field must reverse polarity, but the H field polarity must not change. This means that the current flow on the line behind wave (2) is doubled. When wave (2) reaches the switch, it is again reflected. Poynting's vector for this reflected wave (3) requires that the E field be reversed and the H field again stays constant. Thus, the current in the line after the second reflection is three times the original current. The source supplies a current I until wave (2) returns to the source. After the second reflection, the source must supply $3I$. Obviously, the process repeats itself after each pair of reflections. At the end of the fourth reflection, the voltage source must supply $5I$. The current builds until the voltage source can no longer supply the current, the traces melt or a fuse blows.

A short circuit current develops through a series of wave reflections. The length of time it takes for the current to rise to a given value depends on the characteristic impedance of the line and on the line length. Under ideal conditions, the growth is linear and not exponential.

2.9   WAVES THAT TRANSITION BETWEEN LINES WITH DIFFERENT CHARACTERISTIC IMPEDANCES

When a wave traveling along a transmission line reaches a change in characteristic impedance, some of the energy is reflected and some of it enters the termination. The termination can be a resistor or a second transmission line. If a wave traveling in Z_0 reaches a termination impedance Z_L, the fraction that is reflected is given by

$$\rho = \frac{(Z_L - Z_0)}{(Z_L + Z_0)} \tag{2.5}$$

where ρ is called the *reflection coefficient*. If Z_L equals zero, the reflection coefficient is simply -1. The reflected wave is then the original wave reversed in polarity. If $Z_L = Z_0$, there is no reflection. If Z_0 is large, then the reflection coefficient is near unity meaning that the reflected wave has the same polarity as the arriving wave. In this case, the two waves add together, doubling the voltage at the point of reflection.

N.B.

The reflection coefficient is valid for all wave forms including sine waves and step functions.

The fraction of the wave that continues into the termination impedance is given by

$$\tau = \frac{2Z_L}{(Z_L + Z_0)} \tag{2.6}$$

where τ is called the *transmission coefficient*. If $Z_L = Z_0$ then τ is unity meaning that the transmitted wave is transferred without reflection to the new line or load. If Z_L is high compared to Z_0 then the voltage at the load is double the arriving voltage. This is a voltage doubler without active components.

If a transmission line is terminated in a resistor, the voltage at the resistor is the product of arriving voltage and transmission coefficient. When the termination resistor is higher than the characteristic impedance the voltage at the resistor is shown in Figure 2.7a. When the termination resistor is lower than the characteristic impedance the voltage at the termination resistor is shown in Figure 2.7b.

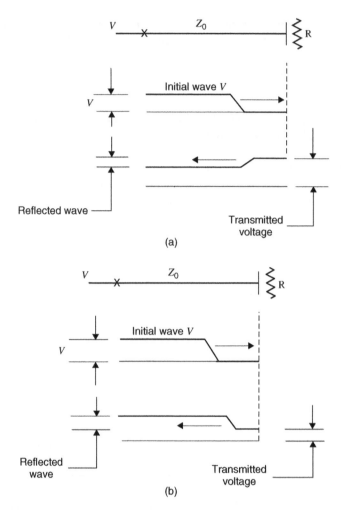

Figure 2.7 (a) The transmitted voltage and reflected wave for $R > Z_0$ and (b) the transmitted voltage and reflected wave for $R < Z_0$.

In a circuit board design there are many places where there are impedance mismatches. Examples might be at a terminating load, at a stub, at a branching transmission line, at an open circuit, or simply at a connector. The nature of the transmission and reflection at a transition in characteristic impedance is shown in Figure 2.8a and b.

There are many transitions that are usually ignored. Examples include the IC die connections, connections to interposer boards, the rotation of field around a mounting pin, transitions through vias, etc.

In circuits where the leading edge takes time to transition, these reflections may not be important. For high clock rates (short rise and fall times), some of these transitions can pose a problem. Examples of problem reflections occur at transmission

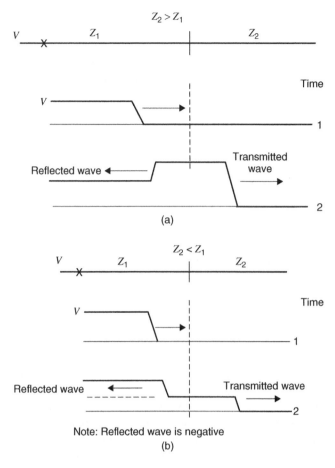

Figure 2.8 (a) Reflection and transmission at a discontinuity where $Z_2 > Z_1$ and (b) reflection and transmission at a discontinuity where $Z_2 < Z_1$.

line branches and stubs. In general, for 50-ohm lines, tests have shown that changes in trace direction or via transitions through a single board layer have a minor effect on wave transmission. Via transitions through multiple layers can cause other problems (see Section 6.19 for a detailed explanation).

N.B.

A transmission line can support any number of waves traveling in both directions at the same time. There is no theoretical limit.

2.10 NONLINEAR TERMINATIONS

When a transmission line is terminated in a high impedance, the reflected wave (2) adds to the forward wave (1). When the full voltage is transmitted down an unterminated line, the reflected wave is clamped (diode action) by the hardware. Since energy cannot be dissipated in a short circuit, a reflected wave propagates back to the source, canceling the voltage. The result is wave action that keeps reflecting between the source and the termination until losses dissipate the excess energy supplied to the line. In general, it is a good practice to design transmission lines so that clamping action does not occur. It is possible that clamping action can fire a four-layer Shockley junction that can result in a catastrophic failure. In some cases, the conduction path through the clamping diode can impact logic operations resulting in logic errors. For these reasons, it is wise to design the logic so that protecting clamps are not used to limit logic levels.

2.11 DISCHARGING A CHARGED OPEN TRANSMISSION LINE

Figure 2.9 shows a length of transmission line that has a characteristic impedance Z of 5 ohm. In this example, the line is initially charged to a voltage of 5 V.

At time $t = 0$, a switch closes and the line is terminated in a resistor R of 50 ohm. At this moment, a step voltage of 4.54 V appears at the load. A step wave (1) of -0.46 V propagates back into the line. At the open end of the line, wave (1) is reflected. The resulting return voltage is 4.08 V. When wave (2) reaches the load, some of the wave energy is transmitted into the load and some is reflected. In this example, the load voltage drops to 3.71 V. After the next round trip of wave motion, the load voltage drops to 3.06 V. The reflection process continues with waves traveling back and forth between the load and the open end of the transmission line. Each time a wave makes a round trip, the energy flowing into the load from the transmission line is reduced. The envelope of decreasing voltage follows an exponential curve with the voltage reduced by the same factor after each round trip. In this example, wave fronts with very short rise and fall times are shown so that the reflection process is clear.

The section of transmission line in Figure 2.9 could represent a capacitor. In circuit theory, the discharge of a capacitor through a resistor follows an exponential curve. The discharge time in the example above depends on the length of the line and the value of the load resistor.

To determine the time constant of the circuit in Figure 2.9, assume that the envelope function is $v = V e^{-t/\tau}$, where τ is the time constant. The derivative of v is

$$\frac{\Delta v}{\Delta t} = -\left(\frac{V}{\tau}\right) e^{-t/\tau} \tag{2.7}$$

At $t = 0$, the derivative is

$$\frac{\Delta V}{\Delta t} = -\left(\frac{V}{\tau}\right). \tag{2.8}$$

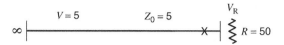

The wave count is shown above each transition

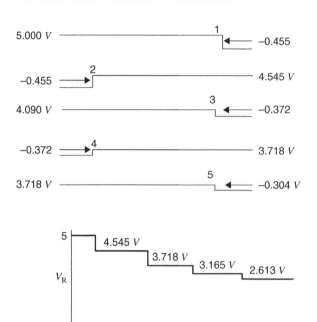

Figure 2.9 The voltage waveform when a transmission line is discharged.

The initial increment of voltage after the wave makes one round trip is approximately $-2VZ/(R + Z)$. If the length of the line is l and the relative dielectric constant is ε_R then the time for a wave to make a round trip is $\Delta t = 2l(\varepsilon_R)/c$, where c is the velocity of light. The ratio of $\Delta V / \Delta t$ at $t = 0$ is then equal to

$$\frac{\Delta V}{\Delta t} = -\left(\frac{2VZ}{(R + Z)}\right) \times \frac{c}{\left(2l\varepsilon_R^{1/2}\right)} \tag{2.9}$$

If we set this equation equal to Equation 2.8, the time constant τ is then

$$\tau = \frac{(R + Z)}{Z} \times \frac{l\varepsilon_R^{1/2}}{c} \tag{2.10}$$

In one time constant, the voltage will sag to $1/e$ or 37% of its initial value. It is interesting to see what the time constant is for a 1.0-cm long 5-ohm line terminated in 50 ohm. If the relative dielectric constant is 400 then the time constant is about 1.5 ns.

The velocity v of waves on a transmission line is given by Equation 2.3 and the characteristic impedance of the line is given by Equation 2.1. Using these equations, where L and C are the inductance and capacitance for the length l, and if we assume $Z \ll R$, the time constant in Equation 2.10 is simply

$$\tau = RC \qquad (2.11)$$

which is the time constant in circuit theory for a parallel capacitor and resistor. This shows quite clearly that the energy being dissipated in the resistor comes from the capacitance of the transmission line. It is interesting to note that the inductance of the line does not appear as a term in Equation 2.11.

N.B.

In a fast circuit, the discharge of a capacitor involves a series of wave reflections. These step changes in voltage represent a rapidly changing field. Circuit theory does not discuss step changes in voltage when a capacitor is discharged, because a step change in voltage requires an infinite current. The problem is circuit theory does not consider a capacitor as a transmission line.

2.12 GROUND/POWER PLANES

Ground and power planes were introduced to digital circuit board design to allow the operation of the circuits at higher clock rates. The improved performance can be related to the tight control of fields associated with the transport of logic signals. It is recognized that the capacitance between ground and power planes stores energy. This capacitance is in parallel with local decoupling capacitors. The energy in these parallel capacitances is available to drive transmission lines. Efforts have been made to increase the stored decoupling energy by using a higher dielectric constant material between the conducting planes. There is a time delay associated with retrieving this stored energy, which requires that decoupling capacitors must still be used. This topic is covered in the next section.

When a trace and a conducting plane are used as a transmission line, the fields that transport signal or energy are tightly confined. The field geometry for a wave is the same whether one of the planes is associated with ground (common) or with the power supply voltage.

> **N.B.**
>
> The average potential on a conducting plane does not affect the field patterns associated with a wave moving between two conductors. The waves in a bathtub are the same whether the bathtub is on the first or second story of a building.

If the conducting plane is floating, the fields are still confined. This is not a recommended practice, as there will be reflections at the plane edges and the plane can become a large radiating surface.

The capacitance in picofarads per square inch of a ground/power plane is given by

$$C = 225 \frac{\varepsilon_R}{t} \tag{2.12}$$

where t is the thickness of the dielectric in mils. This equation assumes that all of the conducting surface contributes to the capacitance. If the dielectric is FR4 then the thinnest practical spacing is about 2 mil.[4,5] The relative dielectric constant of FR4 at 1 GHz is about 3.5.

2.13   THE GROUND AND POWER PLANES AS A TAPERED TRANSMISSION LINE

A ground and power plane can be viewed as a parallel plate capacitor where a via and a trace is used to make a connection to one of the planes. When a connection is made in the middle of a board, the conductor geometry can be viewed as a tapered transmission line (TTL). In this geometry, the characteristic impedance varies with the radial distance from the point of entry. Energy is moved by waves that move radially into or out of the points of contact. When a switch connects a resistive load to the point of entry, the initial wave travels outward radially.

To understand how energy is transported, we will first consider the characteristic impedance of a thin annular ring at a distance r from the point of entry. This geometry is shown in Figure 2.10.

---

[4]A patent is held by Sanmina SCI that provides a basis for a higher capacitance by using a thinner dielectric known as *ZBC*®. The validity of this patent has not been challenged in court. Later, we show that the limitations are not in the capacitance itself but in the time it takes to move stored energy through a practical connection.

[5]Dupont® markets a laminate with a 12 μm polyimide dielectric (0.5 mil). The dielectric constant is about 3.6 over a very wide frequency range. This laminate finds applications in circuit board manufacturing and capacitor manufacturing.

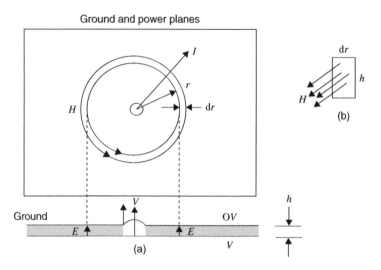

Figure 2.10 (a) A thin annular ring of a tapered transmission line. The H field is circular, the E field goes between the conducting planes, and the current I flows radially in the conducting planes. (b) The H field pattern in the annulus.

The capacitance between two circles of radius r and $r + dr$ and of thickness h is simply

$$C = \frac{\varepsilon A}{h} = \frac{2\pi r \varepsilon \varepsilon_R \, dr}{h} \tag{2.13}$$

where A is the area, h is the spacing between planes, ε is the permittivity of free space, and ε_R is the relative dielectric constant. The inductance of this annular ring can be calculated from the ratio of B flux per unit current. The B flux φ in this annulus is constant and is totally contained. The line integral of H around the annulus is simply $2\pi r H$. This integral must equal the current I. The value of H is

$$H = \frac{I}{2\pi r} \tag{2.14}$$

The induction flux φ is μH times the area enclosed or

$$\varphi = \left(\frac{I}{2\pi r} \right) \times \mu A = \frac{I \, dr \, h \mu}{2\pi r} \tag{2.15}$$

The inductance L is the flux per unit current or

$$L = \frac{dr \, h \mu}{2\pi r} \tag{2.16}$$

The characteristic impedance of this annulus is $(L/C)^{1/2}$. Combining Equations 2.12 and 2.15, the characteristic impedance at a radius r is

$$Z = \left(\frac{h}{2\pi r}\right) \times \left(\frac{\mu}{\varepsilon_R \varepsilon}\right)^{1/2} \tag{2.17}$$

In a vacuum, the ratio $(\mu/\varepsilon)^{1/2}$ is 377 ohm. Assume that the ratio of h to r is unity at the entry point. The value of Z is $377/2\pi = 60$ ohm. If the relative dielectric constant ε_R is 3.5, the impedance at the entry point is approximately 32 ohm. If the spacing between the planes is 10 mil, then at a distance of 0.1 inch from the point of entry, the characteristic impedance drops to 3.2 ohm. At a distance of 3 inches, the characteristic impedance is about 0.1 ohm. Equation 2.17 shows that the characteristic impedance of this type of transmission line is inversely proportional to the radius.

2.14 PULLING ENERGY FROM A TAPERED TRANSMISSION LINE (TTL)

In Section 2.12, we discussed wave action where energy flowed from an ideal voltage source through a short transmission line and was dissipated in a resistor. An increment of energy was transferred for each round trip of the wave in the transmission line. In that discussion, reflections occurred at abrupt changes to the characteristic impedance. In the case of a TTL, the reflection process still takes place but on a continuous basis. A linear change in characteristic impedance results in a continuous reflection process. Of course, there is a reflection at the edges of a board, but this occurs at a much later time. The phenomena we consider takes place before the wave action reaches the edge of the board.

To get some idea of how this TTL works, consider the ideal case where the capacitance of the tapered line is charged to a voltage V. Assume an ideal logic switch and a load of 5 ohm. When the switch closes, if the tapered entry point looks like 50 ohms, the voltage drops to 9.1% of V. We need to find out how long it takes for the voltage at the load to reach 95% of V.

At the moment of switch closure, the wave that propagates into the TTL is -9.909 V. As the wave progresses into the tapered line, the continuous nature of the reflection reduces the magnitude of this wave. Stated another way, as the wave propagates radially, the wave amplitude decreases and the voltage across the leading edge rises. We are interested in the time t it takes for the wave amplitude to diminish so that the voltage across the planes is 95% of V. To obtain this time, the rate of energy transfer from the capacitance to the wave as the wave progresses out on the tapered line must be calculated.

If the wave amplitude progresses until the wave amplitude is -0.95 V, the energy supplied to the wave per unit time depends on the capacitance at that point in its travel. The energy stored in a capacitor between these two voltages is

$$E = \frac{1}{2}C(V^2 - (0.95V)^2) = \frac{1}{2}C(0.0975)V^2 \tag{2.18}$$

Using Equation 2.12 for capacitance, the increment of energy taken from an annular ring at a radius r in a distance dr is given by

$$dE = 0.0975V^2 \frac{\pi \varepsilon_R \varepsilon r \, dr}{h} \tag{2.19}$$

The wave in the capacitor travels at the speed of light divided by the square root of the relative dielectric constant or $v = c/(\varepsilon_R)^{1/2}$. The power W that transfers to the wave at a distance r is given by dE/dt. Dividing both sides of Equation 2.19 by dt and noting that dr/dt is equal to the velocity $v = c/(\varepsilon^{1/2})$, we can write the power as

$$W = 0.095 \frac{V^2 rc\pi\varepsilon(\varepsilon_R)^{1/2}}{h} \tag{2.20}$$

The power required by the load is V^2/R. The value of r where the power supplied to the wave equals the power required by the load is

$$r = \frac{h}{(0.0975c\pi\varepsilon(\varepsilon_R)^{1/2})} \tag{2.21}$$

The time it takes a wave to travel this distance is given by r/v or

$$t = \frac{h}{(0.095R\varepsilon\pi c^2)} \tag{2.22}$$

where c is the velocity of light and $\varepsilon = 8.85 \times 10^{-12}$ F/m.

N.B.

The important thing about Equation 2.22 is that the time t does not depend on the relative dielectric constant.

The time t in Equation 2.22 is the time for the wave to travel out to get energy. The time required to obtain the energy and return to the entry point is double this value or

$$t = \frac{2h}{(0.0975r\varepsilon\pi c^2)} \tag{2.23}$$

If h is 10 mil and $R = 5$ ohm, then the time t is 0.4 ns. This time is too long for logic operating at 1 GHz. This delay is shown in Figure 2.11.

The advantage of using a high dielectric constant is that the wave action stays closer to the point of demand. If several devices take energy from the planes then a higher dielectric constant can reduce cross talk. It may also be useful on circuit boards where the overall dimensions are small.

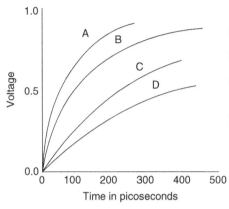

Figure 2.11   Taking energy from the ground/power plane.

Vias and some trace length are needed to provide connections to the ground and power planes. The characteristic impedance of this connection is apt to be greater than 50 ohm. To provide a low impedance connection, assume that the connection was somehow coaxial. To achieve a low impedance connection, the ratio of conductor diameter to conductor spacing would have to be at least 100 : 1. This type of construction is not practical on a typical circuit board. Even if a low characteristic impedance connection could be constructed, how long would the connecting traces be? This argument shows that there are significant physical limitations to moving energy rapidly in or out of a ground/power plane structure.

The waves that propagate between conducting planes reduce in amplitude with increasing radius r. These waves reflect at the board edges and at any discontinuities along the way.

If energy is taken from the two planes near the board edge then the amplitude of the wave at the board edge will be high. Since the reflection at an open edge doubles the voltage, this increases the radiation from the edge. It is good practice to make rapid demands for energy at points that are removed from the perimeter of the board.

N.B.

A wave with an amplitude of 10 mV across a plane spacing of 1 mm has a field intensity of 10 V/m.

2.15   THE ENERGY FLOW THROUGH CASCADED (SERIES) TRANSMISSION LINES

Consider two transmission lines in series as in Figure 2.12. Line 1 has a length d_1, characteristic impedance Z_1, and dielectric constant ε_1. Line 2 has values d_2, Z_2, and

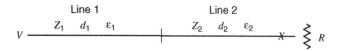

Figure 2.12 Two transmission lines in series with a voltage source. Z is the characteristic impedance, d is the length of line, and ε is the dielectric constant.

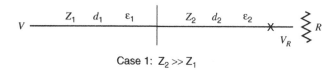

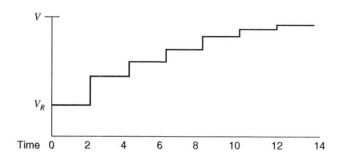

Figure 2.13 Voltage at the load for Case 1 where $Z_1 \ll Z_2$.

ε_2. Initially, the entire line is charged to voltage V. We are interested in the way the voltage builds up in the load R after the switch closes. There are two cases to consider.

Case 1. Line 2 is a short section of high impedance line where $Z_1 \ll Z_2$. As a practical example, consider that line 1 is 10 cm long and has a characteristic impedance of 0.5 ohm and line 2 is 1.0 cm long and has a characteristic impedance of 50 ohm. Assume that the load resistor R is 5 ohm.

When the switch closes, the voltage at the load drops to approximately VR/Z_2. This is also the amplitude of the negative wave that travels back to Z_1. At the interface between the two lines, the initial wave reflects and reverses polarity. When this reflected wave returns to the load, it begins to bring energy to the load. Another way to see this is that the return wave adds an increment of voltage to the load. This wave reflects at the load and starts a wave that makes a second round trip. After each round trip, the voltage at the load rises by an increment. The result of these round trips is a growth in voltage that approaches the power supply voltage exponentially. The voltage at the load is shown in Figure 2.13.

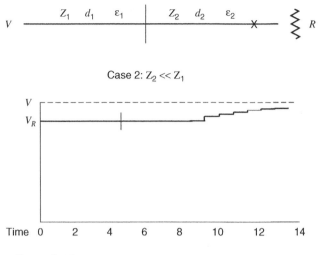

Figure 2.14 Voltage at the load for Case 2 where $Z_2 \ll Z_1$.

The drop in voltage at the load is caused by the very short section of 50-ohm transmission line. This problem corresponds to the problem of drawing energy from a capacitor through a short section of circuit board trace.

Case 2. Line 1 is a short section of high impedance line where $Z_1 \gg Z_2$. Consider that the line 1 is 1.0 cm long and has a characteristic impedance of 50 ohm and line 2 is 10 cm long and has a characteristic impedance of 0.5 ohms.

When the switch closes, the voltage drops a small increment equal to VR/Z_2. The voltage at the load is nearly equal to V. This small increment of voltage is a negative wave that travels to the junction between the two lines where there is a transmission and a reflection. The wave voltage at the interface doubles, transmitting a wave to the voltage source and a reflected wave back toward the load. When the transmitted wave reaches the voltage source, there is another reflection. If line 2 is longer than line 1 then multiple reflections take place on line 1, while a wave makes one round trip on line 2. Each round trip on line 1 sends an increment of energy onto line 2. This string of waves causes the load voltage to approach its final value exponentially. The voltage at the load is shown in Figure 2.14.

These examples clearly show why small sections of high impedance transmission lines must be correctly placed. In Case 1, the voltage sagged, and in Case 2, the voltage remained nearly constant.

N.B.

Transmission lines do not behave like circuit elements in series. The order is important.

2.16 AN ANALYSIS OF CASCADED TRANSMISSION LINES

The following analysis is for Case 2, as shown in Figure 2.14. The time of transit in line 1 is $d_1\varepsilon_1^{1/2}/c$, where c is the velocity of light. Similarly, in line 2, the time of transit is $d_2\varepsilon_2^{1/2}/c$. The ratio of round trip time is

$$\frac{t_2}{t_1} = \frac{d_2}{d_1}\left(\frac{\varepsilon_2}{\varepsilon_1}\right)^{1/2} \tag{2.24}$$

When a load R_L is applied to line 2, a wave propagates on line 2 depending on the ratio of Z_2/R_L. The voltage at the load will stay constant until this wave reflects at line 1 and returns to the load. When the initial wave reaches line 1, a part of the wave is transmitted through to the voltage source. At the voltage source the wave is reversed in direction and amplitude. If line 1 is short then waves will make many round trips between the voltage source and line 2, while a wave makes one round trip on line 2. For each round trip on line 1 there will be some energy supplied to line 2. As these voltage increments reach the load, the load voltage will rise in an exponential manner.

The ratio of energy carried by waves on the two transmission lines is equal to the inverse ratio of characteristic impedances. The ratio of wave round trips is equal to the ratio of transit times as given in Equation 2.2. If we equate the ratio of characteristic impedances to the ratio of transit times, we have an approximate balance in the flow of energy in the two transmission lines. The resulting equation of balance is

$$\frac{d_2}{d_1} = k\left(\frac{\varepsilon_1}{\varepsilon_2}\right)^{1/2}\frac{Z_1}{Z_2} \tag{2.25}$$

where k is about 3. This factor accommodates the fact that the rising voltage at the load approaches its final value in an exponential manner.

It is interesting to use the above equation on an example. Assume d_1 is 1.0 cm $\varepsilon_1 = 4$, $Z_1 = 50$, $\varepsilon_2 = 3600$, and $Z_2 = 1$ then $d_2 = 5$ cm.

If d_2 is shorter than this value then the voltage at the load will sag to a lower value.

If Z_2 represents a decoupling capacitor then this capacitor must have a physical length of 5 cm to be effective. To reduce this length, the capacitor can have a lower characteristic impedance or a higher dielectric constant.

N.B.

The length of the decoupling path is important in supplying a stable flow of energy.

N.B.

It can take many nanoseconds to build up current flow in a 50-ohm line even if it is only 1 cm long. The time depends on the number of round trips waves must make to bring the current to near its expected value.

This build up of current flow in a short section of 50-ohm line illustrates the problem of pulling energy from the ground/power plane capacitance. The characteristic impedance of a typical connection is approximately 50 ohm. It takes nanoseconds before the reflection process can supply current from the planes through this high impedance. If there is a short section of connecting line, the delay is increased further. This is illustrated in curve D of Figure 2.11.

2.17 SERIES (SOURCE) TERMINATING A TRANSMISSION LINE

It is good practice in fast digital logic to use a series line matching resistor at the logic source. This circuit is shown in Figure 2.15.

When the logic transitions to a voltage V, the voltage that propagates down the line is $V/2$. When this wave reaches the open end of the line the reflected wave is

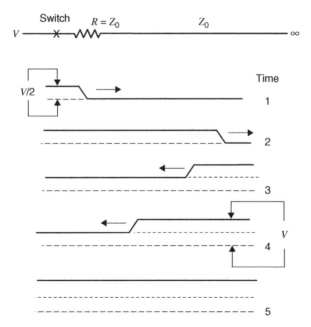

Figure 2.15 A series (source) terminated transmission line.

equal to $V/2$. The reflected wave cancels the current and raises the voltage to V along the length of the line. When the reflected wave reaches the source, the voltage is V at both ends of the line and the current is zero. At this time, all wave action stops and no further energy is supplied to the line.

There are several benefits to this approach. The current supplied by the logic is one-half the value for the end (parallel) termination case. Radiation levels are lower, because the wave voltage is half the driving voltage. The energy supplied to the transmission line equals the energy stored in its capacitance plus the energy dissipated in the matching resistor during the round trip time. Several lines may be driven in parallel when each line is supplied with a line matching resistor.

2.18 PARALLEL (SHUNT) TERMINATIONS

In high speed circuits, parallel transmission line terminations can be useful. This termination is placed at the end of the transmission line as shown in Figure 2.16.

Since there is no reflection at the ends of the lines, line current continues to flow as long as there is a signal. Because of the extra power consumption, this technique is usually limited to low voltage logic. In general, this logic design approach has a higher noise margin than the series termination case.

In series terminated circuits, the clock signals are usually timed so that logic signals have time to reflect and return to the source. This timing guarantees that all logic connections made along the transmission path will receive the doubled voltage before the clock signal arrives.

On long lines with all the loads at the end of the lines, parallel terminations can be effective. The clock signal can arrive anytime after the logic signal reaches the termination. On a long line, compensation for line loss can be made by increasing the value of the terminating resistor. The reflection that results adds to the signal, thus, reducing the

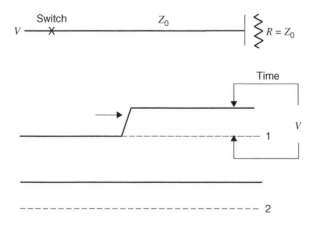

Figure 2.16   A parallel terminated transmission line.

chances of error. Parallel termination has the drawback that energy must be supplied to the line as long as the logic signals are present. If the transmissions are infrequent and the logic levels are at zero, most of the time, this can be a very effective approach.

When an external cable brings a balanced logic signal to a circuit board, the cable must be terminated in its characteristic impedance. If the cable impedance is 100 ohm and if the individual trace impedances are 50 ohm then the cable impedance is matched at the connection. In this approach, the traces must be routed independently (spaced apart) and connected to the proper logic pins. Each trace is then properly terminated (typically 50 ohm). If the cable conductors are connected to a 100-ohm transmission line on the board, the cable would in effect be double terminated. This problem is discussed at the end of Section 3.5.

N.B.

Terminating a balanced cable to a circuit board may involve supplying a bias voltage. The bias resistors become a part of the termination impedance.

The parallel termination of stripline poses a special problem. Assume that the stripline is between a ground and power plane. The transmission line energy can be thought of as traveling in two paths. The first path is between the strip line and the ground plane, and the second path is between the transmission line and the power plane. The correct approach for terminating symmetric stripline is to terminate each energy path separately. If the characteristic impedance is 50 ohm then the correct parallel termination is one 100 ohm resistor to the ground plane and one to the power plane. These connections would be at the power and ground pins of the receiving logic. With this approach, it is easy to see where the return current flows.

If one 50-ohm terminating resistor is used, there are several problems. The 50-ohm resistor is a mismatch for just one of the transmission waves. This means that there will be a reflection at the resistor on this half of the stripline transmission. The second energy path sees no terminating resistor. Some of the energy at this mismatch will propagate and reflect into the ground/power plane. The energy will also transmit and reflect at any nearby decoupling capacitors. It is easy to see that a single parallel terminating resistor can cause many unexpected reflections. Problems may not occur if the rise times are greater than 1 ns.

If the stripline is run between two ground planes then two parallel terminating resistors are still required. If one resistor is used, the reflections will involve connecting vias instead of decoupling capacitors.

N.B.

The use of a series termination in stripline avoids the issues presented above.

2.19   STUBS

A stub is a short section of transmission line, which is parallel connected to a longer or main transmission line. This configuration is shown in Figure 2.17.

The stub allows a logic signal to be parallel connected to two different points in the circuit. In general, stubs are not terminated in a matching impedance. Stubs will often terminate at a gate on an IC. The termination will usually be slightly capacitive, but in our analysis, we will consider it as an open circuit.

N.B.

Stubs are sometimes referred to as *loads*, although they are not terminated.

N.B.

If a line plus its stub are shorter than one-quarter the distance a wave travels in one rise time then the delays caused by the stub will not be a problem.

At a clock rate of 1 GHz, the rise time must be less than 100 ps. The distance a wave travels in 100 ps is 1.5 cm. One-quarter of this distance is 0.375 cm. This means that in any realizable gigahertz circuitry, stubs should be avoided.

In Figure 2.18 that follows, the logic level is a nominal 1.0 V. In practical applications, this voltage must be scaled for the logic that is used. High speed logic voltage can be as low as 800 mV or as high as 3.5 V.

Case 1. The driver source impedance is zero and the main transmission line is series terminated in its characteristic impedance. Figure 2.18 shows wave forms for a stub at the midpoint of a line that is 20 ps long, where the stub is 10 ps long.

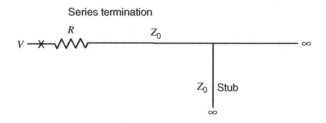

Figure 2.17 A stub added to a transmission line.

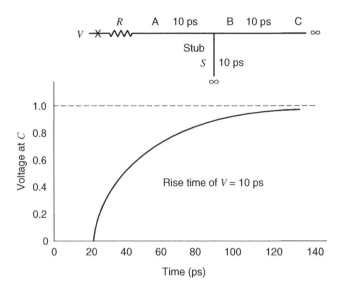

Figure 2.18 The signals on a serial terminated transmission line with a stub at its midpoint.

The initial wave that arrives at the stub divides into three equal waves. The first wave takes the path AB and arrives at the main line termination as 0.333 V. The next increase in voltage follows for a wave that takes the path ASSB. This wave adds 0.2 V to the termination after a delay of 40 ps. The wave taking the path AAAB makes a round trip to the stub and then continues to the termination. The path for the next wave to reach the load is ASSAAB or a delay of 60 ps. Two round trips to the source of voltage is a delay of 120 ps. If the initial rise time is added to this figure, it is easy to see how the total rise time is affected. The increased rise time is a function of where the stub is placed along the line and the length of the stub.

Case 2. The main line is parallel terminated at the logic termination. The terminating resistor matches the characteristic impedance of the transmission line. The waves that are reflected and returned to the main line are absorbed in this termination. The effect on rise time is about the same as Case 1.

N.B.

The increased rise time caused by a stub must be included in designing a logic circuit.

If an unterminated line has an unterminated stub, the reflections can, in some cases, become more than double the initial voltage. The rise time rule must be applied to the length of the line plus the length of the stub. It is interesting to place some numbers on the delays caused by stubs. Assume that the velocity of a wave is 0.015 cm/ps.

A wave on a stub that is 1.5 cm long has a round trip time of 200 ps. More than one round trip on the stub is usually involved. The signal delay must be considered when timing clock signals. If the propagation time is 100 ps and the initial rise time is 30 ps, the delay if there are two round trips would be 530 ps. If a stub is very short then waves can make many round trips on the stub, while one round trip is made on the main line. This has the effect of filling the stub with energy, thus, reducing the delay caused by the stub.

N.B.

The delays caused by stubs can be significant. They must be included in designing a logic circuit.

2.20 DECOUPLING CAPACITOR AS A STUB

Consider a decoupling capacitor placed between a voltage source, an IC, and a driven logic trace. The capacitor can be viewed as a low impedance transmission line stub. When the IC switches, the demand for energy moves on four transmission lines. These are the logic line, the power line, the capacitor as a stub, and the connections to the stub. The waveform at the receiving logic depends on the reflections and transmissions that take place for this configuration. A little thought will show that dozens of reflections are involved before the voltage at the load stabilizes. If the IC is already driving other logic lines, the wave action is even more complicated.

2.21 TRANSMISSION LINE NETWORKS

A transmission line network includes the conductor pairs that supply or carry energy when a logic transition occurs. When a logic switch closes, waves travel over the entire logic structure. The nearby transmission lines that supply the first energy control the sag in voltage. These lines might be considered the network of interest. This network might include the following:

1. the power path connecting power to the IC (bonds, pads, traces, pins, vias and legs);
2. stubs or branches on the signal path;
3. all traces that are connected to logic 1 at the IC;
4. the ground/power plane;
5. the transmission lines that connect the decoupling capacitors;
6. the transmission lines internal to the decoupling capacitors.

The waves that transition in the network cause the voltage to sag, as energy is moved in the network. In a typical process involving just three transmission lines, 20 or 30 reflections and transmissions can take place before a useful signal might arrive at a logic gate. In practice, the details of these multiple reflections and transmissions are obscured by rise time phenomena and the fact that waves from earlier logic transitions are still creating signals.

The program outlined in the next section ignores rise time and line losses. All interconnected transmission lines are initially set at the power supply voltage. At $t = 0$, a first wave is generated by a switch closure, and it is assigned the wave number $m = 1$. Each segment that makes up the network is assigned a unique location number. Consider a wave number 7 in segment 4. The voltage of wave 7 is WV(7). The location of wave 7 is LOC(7) = 4. The direction of WV(7) is given by DIR(7), which can be $+1$ or -1. The transit time for every line segment is one of the initial parameters. The arrival time for WV(7) is the wave start time plus the transit time. The arrival time for WV(7) is given by TM(7).

When the first wave reaches the end of its segment, new waves are generated. These wave voltages are perhaps WV(2) and WV(3). If WV(2) is the transmission of WV(1) into segment 2 then we can assign LOC(2) = 2 and DIR(2) = 1. If wave WV(3) is reflected back on to line 1 then LOC(3) = 1 and DIR(3) = -1. The wave voltage WV(2) = WV(1) $\times$ TTL12 and wave voltage WV(3) = WV(1) $\times$ RTL21. The term TTL12 is read, transmission from transmission line 1 to 2. The term RTL21 is read, reflection from line 2 back into line 1. After this transition, wave W(1) is no longer active and a flag FLG(1) is set to 1.

The program must provide reflection and transmission coefficients for both ends of every transmission line segment. The reflection coefficient for a zero impedance point (voltage source) is -1. The reflection coefficient for an open circuit is $+1$. The transmission and reflection coefficients at a simple transition in characteristic impedance are discussed earlier in this chapter. If a line branches into two lines then six coefficients are required for that junction.

The program starts by incrementing time. The active waves are examined one at a time in a program loop using a counter n. If the program time is greater than an arrival time then a wave has reached its segment end. The program then generates new waves and the old wave is flagged (it is now inactive). The voltage on each segment at any time is the initial voltage plus the sum of all the flagged waves that have traveled along that segment.

2.22 THE NETWORK PROGRAM

The following program outline can accommodate a large group of interconnected transmission lines (network). The basic program provides for one logic transition at $t = 0$. The program can be modified to allow for multiple logic transitions occurring at different times. The increment of time should be small enough that waves are detected, as they reach the ends of each segment.

The program below stops after 300 waves have been assigned. This number is arbitrary and can be changed to fit the problem being solved.

PROGRAM OUTLINE

Draw the network. Assign a number to each transmission line segment.

100 Provide initial voltages, termination and characteristic impedances, transit times.

120 Provide reflection and transmission coefficients for every wave transition, reflection, and termination. Let n be the program counter. Set $n = 1$. Let m be the wave counter. Set $m = 1$.

130 Set WV(1) to a voltage, a location LOC(1), a time of arrival TM(1) in picoseconds and a direction DIR(1). This is the initial wave that occurs after a logic switch closes.

140 Increment t (time) in picoseconds.

150 Increment n. IF $n > 300$ then END. Note that n cycles through all m wave numbers.

160 If DIR(n) = 0 then set $n = 0$: (End of assigned waves) GOTO print routine 400.

170 IF FLG(n) = 1 THEN 150 ELSE 180 'This wave has completed its transmission'.

180 IF $t >$ TM(n) THEN 200 ELSE 150. 'A wave has reached a termination or branch point'.

200 Increment m. 'Calculate each WV(m) at the interface defined by LOC($m - 1$)' and DIR($m - 1$). Assign values to DIR(m), TM(m), LOC(m). For each wave generated at this interface increment m and assign new values to WV(m), DIR(m), TM(m), and LOC(m)'.

210 GOTO 150.

400 FOR $j = 1$ TO $m - 1$.

420 IF FLG(j) = 1 add wave voltage to initial voltage for each line segment.

440 NEXT j.

450 PRINT t, m, segment voltages. GOTO 140.

The printout will show voltages on each segment as a function of time.

2.23 MEASURING CHARACTERISTIC IMPEDANCE

A time domain reflectometer can be used to measure the characteristic impedance of a transmission line. The line is driven with a step function through a known matching resistor. If the transmitted voltage is reduced by a factor of 2, the characteristic impedance is equal to the resistor value. The voltage is sensed after the initial rise time. Obviously, measurements made on actual traces allows for very little experimentation.

An analog method can be used to measure the characteristic impedance of many transmission line geometries. The method makes use of the fact that characteristic impedance involves the ratio of conductor dimensions. For example, two 5-mil wide traces that are 2 mil thick that are spaced 5 mil apart has the same characteristic impedance as two bars of metal 0.5 in wide and 1/16 in thick, which are spaced 0.5 in apart. The bars of metal can be laid out on the surface of a test bench. The capacitance of this model can be measured by using simple test equipment. The characteristic impedance is related to the capacitance by Equations 2.3 and 2.4. The capacitance of the model is the capacitance of the traces multiplied by the model scale factor. In the example above, this factor is 100.

The conductors used in a model must be long enough to provide a useful capacitance. In the example above, a length of 1 m will yield a capacitance of about 50 pF. To measure this capacitance, a function generator that supplies triangle waves can drive the capacitance through a 100-ohm series resistor. The voltage can be set to change 20 V in 0.5 μs. Here dV/dt equals 40 V/μs. The current in the capacitance is $C dV/dt$ or ± 200 μA. The voltage across the resistor in this example is ± 20 mV. The characteristic impedance is $1/cC$, where c is the velocity of the wave in meters per microseconds and C is the capacitance in microfarads per meter. In this example, $Z = 1/300 \cdot 50 \cdot 10^{-6} = 66.6$ ohm.

For traces over a ground plane, the electric field concentrates in the dielectric. This means that most of the energy travels in the dielectric. For side-by-side traces over a dielectric, the electric field divides between the air and the dielectric. Note that the energy density in the dielectric is reduced by the relative dielectric constant. This means that more of the energy travels in air, and this energy travels faster. A measure of the capacitance does not relate to how the wave action divides or to the fact that the wave moves slower in the dielectric. Note that the time domain reflectometer method of measurement does not consider the way the energy divides between air and the dielectric.

This modeling method can also be used to measure leakage capacitance to a nearby trace. The measurement is made by connecting the nearby trace to common through a 100-ohm resistor and noting the voltage across the resistor. When the signal line is driven, some of the electric flux terminates on the nearby trace. This changing flux induces the current that is measured. The leakage capacitance determined in the model must be reduced by the scale factor to obtain the capacitance on an actual circuit board. In the example above, the scale factor is 100. If the mutual capacitance of the model is 3 pF, the mutual capacitance to a centimeter of trace is 0.03 pF. It is easy to see that this modeling method makes it possible to measure very small capacitances.

GLOSSARY

Cascaded transmission lines (Section 2.16): Transmission lines in series.

Characteristic impedance (Section 2.4): The ratio between voltage and current associated with a wave traveling on a transmission line. The square root of the ratio of inductance to capacitance per unit length of a transmission line or $(L/C)^{1/2}$.

Clamping action (Section 2.10): Diodes that are placed on integrated circuit connections to protect the circuit from over or under voltages. It is wise to design logic circuits so this diode action is not used.

Common-mode signal: The average voltage on a group of conductors measured with respect to the receiving common conductor (ground).

Decoupling capacitor (Section 2.12): A capacitor that stores field energy that is mounted near or at a component to supply transmission line energy when logic switching takes place.

Embedded microstrip (Section 2.3): Traces on the outer layer of a circuit board that are embedded in a dielectric. This practice has many benefits in short rise time logic.

Even mode (Section 2.3): A signal that is common to a pair of signal traces (differential pair). For differential signaling, the even-mode signal is an unwanted signal often called the *common-mode signal*. The receiving logic is designed to reject this part of the incoming signal and respond to the normal mode (odd mode) part of the signal.

Exponential (Section 2.11): A process that involves factors of $e^{\alpha t}$, where e is the base of natural logarithms and t is the time.

Exponential growth: Any process where the growth rate is proportional to the amount present.

Exponential decay: Any process where the amount present falls off proportional to the amount present.

Ground (Section 2.3): The conductors on a logic circuit board that are at the reference potential. In 5-V logic, it is the conductor at 0 V.

Ground/power planes (Section 2.3): Conducting planes that are used to bring power and power return to circuit elements on a circuit board. These planes are also used as one side of logic transmission paths. These planes can be islands of copper or the flooding of copper on a dielectric surface. These planes usually start as a copper foil laminated to a dielectric.

Impedance mismatch (Section 2.7): A transmission line termination that is not equal to the characteristic impedance of the line. The termination can be a resistor or another transmission line.

Line: A transmission line or trace.

Leakage capacitance: Parasitic capacitance; mutual capacitance.

Lumped capacitance (Section 2.7): The total capacitance of a transmission line; the representation of a transmission line using shunt capacitors.

Lumped inductance (Section 2.7): The total inductance of a transmission line or circuit; the representation of a transmission line in terms of series inductors. This description implies that the magnetic field is totally confined.

Microstrip (Section 2.3): Traces on outer layers of a circuit board.

Odd-mode logic (Section 2.3): A logic signal carried by a differential trace pair where trace 1 has a logic 1 when trace 2 has a logic 0 and vice versa.

Open circuit (Section 2.7): The lack of termination. A trace that terminates on a logic gate without a parallel termination, resistor is said to be terminated in an open circuit even though the gate has some capacitance.

Parallel terminations (Sections 2.6 and 2.18): A terminating resistor (load) at the end of a transmission line. Sometimes, called a *shunt termination*.

Shunt termination: See Parallel termination

Source termination, Series termination: The line termination resistor is placed at the line driver.

Step voltage (Sections 2.1 and 2.6): An ideal step voltage is a voltage that transitions between two values in zero time. All practical step voltages have a finite rise or fall time.

Stub (Section 2.19): A short section of transmission line that connects to a main transmission line somewhere along its length. Stubs are usually not terminated. In gigahertz logic, they should be avoided.

Reflection coefficient (Section 2.9): The fraction of an incoming wave that is reflected and returned when a transmission line is terminated in a mismatch.

Stripline (Section 2.3): Traces on inner layers of a circuit board.

Tapered transmission line (TTL) (Section 2.13): Two conducting planes separated by a dielectric where a coaxial connection is made at a central point. The ground/power planes.

Termination (Section 2.2): The circuit that connects to the end of a transmission line. It can be a resistor, a second transmission line, or a logic gate.

Time constant (Section 2.11): In exponential decay, the time it takes to fall to 37% of the initial value. For a capacitor discharged by a resistor, the time constant is simply RC, where t is in microseconds, C is in microfarads, and R is in ohms. See exponential decay.

Traces (Section 2.3): Thin conducting strips of copper. These strips can be formed by plating copper on a dielectric laminate or by etching a copper foil bonded to a dielectric laminate. Traces are circuit conductors carrying signals or power on a circuit board. Traces located on the outer layers of a circuit board are called *microstrip*. Traces on an inner layer are called *stripline*.

Transmission coefficient (Section 2.9): The fraction of arriving signal that is transmitted into a terminating transmission line or into a terminating component. If the characteristic impedance of the line matches the load then the transmission coefficient is 1.

Transmission line termination (Section 2.2): A resistor equal to the characteristic impedance of a transmission line inserted in series with the line driver (series termination) or placed across the line at the line termination (parallel termination). A second transmission line can also be a termination.

Via (Section 2.14): A conducting path that connects traces located on different layers of a printed circuit board. Blind vias do not go through the board. Buried vias connect conductors on inner layers. Vias are usually plated through holes.

Wave velocity (Section 2.5): $(LC)^{-1/2}$ The velocity of the leading edge in transmission lines. The wave velocity is equal to the speed of light divided by the square root of the relative dielectric constant. If the relative dielectric constant is 4 then the wave velocity is 15 cm/ns. Distortions of the leading edge will change the apparent velocity.

3

RADIATION AND INTERFERENCE COUPLING

3.1 INTRODUCTION

The waves found in logic are step functions and the radiating geometries are certainly not simple dipoles. Most of the literature that is available on radiation discusses sinusoidal wave patterns from antenna or wave guide type structures. Circuit board radiation does not result from a recognized structure, and the wave forms are not sinusoidal. To discuss radiation and field coupling, we will use the accepted terminology although it may not be a good fit. For example, the term *wave impedance* applies to sine waves and not step functions.

Interference on a circuit board can result from external fields in the environment or from fields that originate on the circuit board. Cables that interconnect circuit boards or hardware can carry interference fields into the circuitry. This is sometimes called *conducted interference*. When these fields couple into logic signal paths, they take up a part of the available noise budget.

Digital Circuit Boards: Mach 1 GHz, First Edition. Ralph Morrison.
© 2012 John Wiley & Sons, Inc. Published 2012 by John Wiley & Sons, Inc.

3.2   THE NATURE OF FIELDS IN LOGIC STRUCTURES

The characteristic impedance of a transmission line depends on conductor geometry. It defines the ratio between the E and H field intensities[1] that propagate along that line. On a circuit board, the very close spacing between traces and from traces to conducting planes limits the field energy that leaves (radiates from) any one transmission line. For traces that are between conducting planes, there is essentially no board radiation. On a typical outer layer, there may be hundreds of transmission lines in operation at the same time. If a small fraction of the transported energy leaves each transmission line, the result can be a significant radiated field.

3.3   CLASSICAL RADIATION

The fields associated with sine wave voltages and currents provide insight into the radiation process. We have already seen that any electrical activity involves both fields. In ac circuit theory, the energy associated with a magnetic field in an inductor is returned to the circuit twice per cycle. The same process takes place with a capacitor except that this energy is in an electric field. In an ideal component, very little of the field energy leaves the component space. In high frequency circuits the component space is not that well defined and the fields are also not well confined. In early rf circuitry components were often placed in metal cans to avoid cross coupling.

At low frequencies (below 100 kHz), the field energy associated with electrical activity moves into position without an apparent delay. The field moves out from the circuit at the speed of light. To begin our discussion of radiation, consider a sine wave magnetic field associated with an inductance. As the frequency is increased, a component of the magnetic field cannot return to the circuit in phase with the current generating the field. This occurs because the field travels out and back at the speed of light. This is the magnetic field component of the radiated field. A changing magnetic field is always associated with a changing electric field. This is one of Maxwell's equations. This electric field also moves out into position at the speed of light. As the frequency is increased, a component of the electric field cannot return to the circuit in phase with the generating voltage. These two fields, taken together, leave the circuit as radiation.

The wave character of a field near a radiator is given by the ratio of E and H field intensities. Near the radiator, the field is a composite of radiating and returning field energy. Near loops, the magnetic field intensity is high and the ratio between E and H field intensities is low. Near radiating dipoles, the electric field intensity is high and the ratio of E to H field intensity is also high. Since the ratio of E to H has units of ohms, it is usual to describe the field character of waves in terms of wave impedance. The wave impedance near loops is low. The wave impedance near isolated conductors (dipoles) is high. The radiating intensities for a loop and for a dipole are shown in Figure 3.1.

---

[1] The characteristic impedance is the ratio of E to H field intensities at all points in the space between the transmission line conductors. Each wave that is in motion must be considered separately.

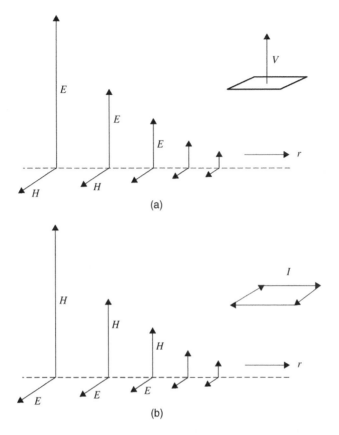

Figure 3.1 Radiated field intensities from a dipole and from a conducting loop. (a) The radiated field near a half dipole; (b) the radiated field near a current loop.

N.B.

The wave impedance in space far removed from a radiating source is constant at 377 ohm.

3.4 RADIATION FROM STEP FUNCTION WAVES

In logic structure, the waves are step functions and not sinusoids. A spectrum of frequencies with different amplitudes makes up this wave, and the radiation at each frequency is different. The higher frequency components are more efficient radiators. It is difficult to relate this type of signal to wave impedance. We can assume that the fields that are generated by loops will have a high H field content. It is correct to say that loops generate fields that have a low wave impedance character.

N.B.

The E field has units of volts per meter, the H field has units of amperes per meter.

The ratio of E/H has units of volts per ampere or simply ohms. There is no available way to measure wave impedance directly.

It is interesting to note that characteristic impedance and wave impedance are both measured as the ratio of field intensities. In transmission lines, characteristic impedances are usually below 300 ohm. Near dipoles, where the H field is very small, the wave impedance can be well over 5000 ohm.

At a distance from a sinusoidal radiator, the ratio of E field to H field intensity in space is 377 ohm. These waves are referred to as *plane waves*. The distance from a sinusoidal radiator where the wave is considered to be a plane wave is $\lambda/2\pi$, where λ is the wavelength of the sine wave signal. This length is called the *near-field/far-field interface distance*. Beyond this distance, both fields attenuate proportional to the distance, and the ratio of field intensities is constant at 377 ohm. Near loops, the wave impedance can be a few ohms. This type of field is called an *induction field*. The fields near a radiating dipole are of high impedance.

The impedance of a field says a lot about how it can be shielded. In general, high impedance fields can be shielded by a sheet of conducting material. Utility power generated fields are usually of low impedance, and they are hard to shield. On a circuit board with logic traces where the radiation comes from loops, the wave impedances are considered to be low.

For nonsinusoidal fields, it is useful to assign a single frequency for an analysis. This frequency is equal to $1/\pi\tau_r$, where τ_r is the rise time of the waveform.[2] This frequency can be called the *rise-time frequency*. This frequency is sometimes called the *maximum pulse frequency*. To obtain a measure of how a circuit might respond to a pulse or step function, this single sinusoid can be used as an input. The peak value of the sinusoid selected should be equal to the peak value of the step function or pulse. The response of the circuit or system to this sine wave signal can be an indicator of what to be expected when an actual transient takes place. For transmission lines, the wave impedance near the line can be considered the characteristic impedance of that line. Typically, this impedance is about 50 ohms.[3] For logic with a rise time of 100 ps, the rise-time frequency is about 3 GHz. The near-field/far-field interface distance using this frequency is about 2 cm. Beyond this distance, the wave impedance

---

[2]Both rise and fall times must be considered. Use the smaller number. Rise time is usually specified as the time between 10% and 90% points. In digital circuit analysis, it is safer to consider the time between 20% and 80% points.

[3]Trace impedances are about 50 ohm compared to a free space of 377 ohm. This would imply that the coupling between adjacent traces is dominated by induction. It turns out that for stacked traces, the cross coupling is mainly capacitive (Section 3.8).

for a wave is 377 ohm. This means that it is relatively easy to shield fast logic if it is needed.

N.B.

Electromagnetic fields transport all energy on a circuit board. There is no way to tell if a wave is carrying information or it is supplying energy to an IC.

The radiation from a circuit board comes from identifiable loop areas. The loops include the spacing between the traces and their signal returns, connections to decoupling capacitors, the connections to components, and the loop areas associated with vias. The individual loop areas are small, but their numbers can be large. It could be argued that the orientation of these loops is random, and this limits the total radiation. In any worst case analysis, it is a good idea to consider all radiation as additive.

An approximate equation for calculating radiation from a single loop is

$$E = 6 \times 10^{-3} \frac{IAf^2}{r} \tag{3.1}$$

where E is in decibel-microvolts per meter, I is in milliamperes, A is the loop area in squared centimeters, r is the distance to the radiator in meters, and f is the rise-time frequency in megahertz. This equation provides an rms measure of field strength without regard to waveform. For an entire board the total value for E is the sum of the radiation from each loop. Different logic families will require different rise-time frequencies. In this situation, the total radiation is the rms sum for each frequency group. Note that the E field falls off proportional to distance not as the square. The energy density falls off as the square of distance.

The loop areas formed by traces on outer layers must be considered first. The traces between conducting planes do not radiate directly into space. If inner traces do couple to energy, they can radiate this energy when they cross to an outer layer (see the discussion in Section 3.11).

Interference fields that are created by other circuits or external hardware can couple to entering cables. This interference can be carried to all parts of the board, including the space between conducting planes. Because transmission lines are two-way streets, connecting cables can also carry interference to the outside world.

N.B.

All transmission lines can carry energy in both directions.

Coupling processes occur in areas where waves are in transition. Even though there is energy in motion behind a wave front, there is no coupling where the wave is not changing. If the length of a transmission line is equal to the distance, a wave travels during its rise or fall time, then all points on the trace will radiate at the same time. Note that in Equation 3.1, the loop area is limited to the area where the wave is in transition. For a 100-ps rise time, the loop length is only 3.5 cm.

N.B.

Cross coupling can be high for stacked traces. The wider the traces the greater the cross coupling. When there are two trace layers between conducting planes, it is preferred that the traces on layer 1 run perpendicular to the traces on layer 2. When there are parallel runs, an analysis of signal integrity is recommended. Staggering the traces will reduce the trace-to-trace capacitance.

3.5 COMMON MODE AND NORMAL MODE

To appreciate the meaning of these terms, it is worth discussing the rejection of common-mode interference in analog circuits (below 100 kHz). The circuit geometry is very similar to the arrangements found in logic transport between different pieces of hardware. In analog signal processing, the signal of interest is often generated at a remote point and carried to conditioning electronics over a long cable. The zero or reference potential for the signal is often carried on a shield conductor G_1 that is grounded near the signal source. The circuits that amplify the signal are associated with a second ground G_2 located at the point of amplification.

The potential difference that can be measured between G_1 and G_2 at G_2 is generated by fields in the area of cable routing. Consider a cable routed between two pieces of hardware. Fields in the area cross the loop formed by this cable and nearby conducting structures (grounds). These fields are often related to utility power, but they can include fields from radio and television transmitters, as well as digital circuitry. Signals that are observed between G_1 and G_2 at the receiving end of the cable will also appear between G_1 and all the signal leads in the cable.[4] This common signal must be attenuated or filtered so that the signal of interest (signal difference) can be amplified. This average or common signal is called a *common-mode signal*. For low frequency analog signals, these circuits are called *differential amplifiers*. This type of linear amplifier conditions (amplifies) the signal of interest (normal-mode signal) and attenuates the common-mode signals. In analog instrumentation, the common conductors are not usually connected together at the interface.

For signal conditioning above 100 kHz, the shield of a connecting cable should be grounded at both ends. This is the case when digital signals are transmitted over some distance. Consider Figure 3.2 where an external cable carries signals between devices.

---

[4]This assumes that G_1 and G_2 are not tied together near the point of measure.

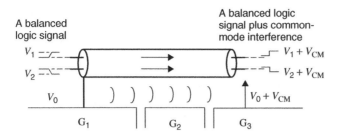

Figure 3.2 A long cable between devices. Note: The shield may not be present.

Fields in the area will use the cable shield as one side of a transmission line. The return path can be racks, shields, other cables, or earth. The protection that could be provided by a close conductive plane is absent. This means that external fields can couple to the loop area formed by the cable and its return path. When this cable is connected to a circuit board, the arriving energy becomes a part of the general ambient. Some of the energy is reflected and some of it follows conductors and traces to their terminations on the circuit board. The average interference signal present on a group of logic lines, when measured with respect to a local common or ground, is called a *common-mode signal*. In digital parlance, this signal adds even-mode interference to odd-mode logic. This interference can use up available noise budget.

Definition: *Common-Mode Signal*. The average interfering signal on a group of conductors measured with respect to a local reference conductor.

Definition: *Normal-Mode Signal*. The signal difference of potential between a pair of conductors. A normal-mode signal is also called the *signal of interest*.

N.B.

The average signal on a group of conductors does not mean that it should be rejected. Consider a single-phase power circuit. The average voltage is 57 V. On a 5-V power supply, the average voltage is 2.5 V. These voltages are needed for circuit operation.

Cables often carry balanced logic lines through connector pins and then onto connecting traces. These traces may be stripline or microstrip. The receiving traces and their termination require careful attention. A typical common-mode rejection circuit is shown in Figure 3.3. Note that this circuit can accommodate interference that drives the input leads negative.

When a logic stream must cross between different pieces of hardware, external fields in the area will couple common-mode interference into the signal path. If the signal is balanced and if the cable pair is balanced, any coupled common-mode interference can be rejected by the electronics. A balanced signal pair means that the

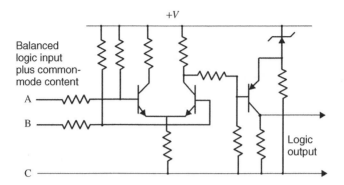

Figure 3.3   A differential logic circuit.

average signal is a constant. A balanced cable pair means that interference couples equally to both lines. A balanced logic signal requires that when line 1 is at a logic 1, line 2 is at logic 0 and vice versa. A balanced cable pair implies that an interfering field will add the same voltage to both logic lines. If the logic is 0 and 5 V, the interference at one moment might add a half volt to each line. At that moment, the logic signals are 0.5 and 5.5 V. The logic only considers the difference that is still just 5 V.

N.B.

Differential logic receivers can accommodate a loss of signal in the cable.

N.B.

A balanced line does not mean that interference will couple equally to both lines. Any interference that couples to just one line creates normal-mode interference that cannot be rejected by electronics. This means that the designer must be careful in routing traces near lines carrying balanced signals.

The termination of an external cable carrying a balanced signal requires some explanation. For short rise-time applications, the line termination should take place at the receiving logic. In many slower applications, the termination resistors can be located at the connector. Let us treat the short rise-time case first.

The balanced line impedance of a cable is typically 100 ohm. This impedance should be matched where the cable enters the board. If the cable connects to two 50-ohm traces then the two trace impedances appear in series, and this matches the

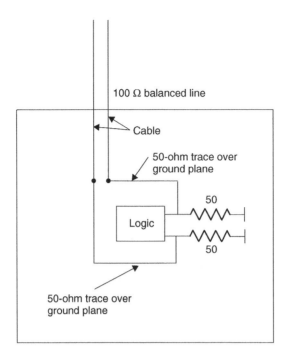

Figure 3.4 The termination of a balanced transmission line on a circuit board.

cable impedance. These traces must not be closely spaced or the cable will be doubly terminated. In effect each trace as a transmission line carries half of the energy flowing in the cable. At the electronics, the two transmission paths are terminated in 50-ohm resistors (Fig. 3.4).

If the cable terminates on parallel board traces that form a 100-ohm transmission line, there are really three energy paths, namely, the path from each line to ground and the path between traces. Each path must be terminated or there will be reflections. If the ground plane is removed then there is only one energy path and one terminating resistor.

If the terminating resistors are placed at the cable connector, the traces from the connector to the IC form a stub. If this stub is shorter than one-quarter the distance a wave travels during the rise time then there will be no problem.

N.B.

The output of the balanced receiving logic circuit is a single-ended logic line.

Cross talk has the effect of introducing jitter into logic. Ideally, if the common-mode interference affects each lead equally, the effect is balanced out and the logic transition time is not affected by the interference. Traces carrying balanced signals should not be routed parallel to long traces carrying unbalanced signals. If one of the traces should couple to interference this is odd mode coupling. This mode is not rejected in the electronics.

3.6 THE RADIATION PATTERN ALONG A TRANSMISSION LINE

The energy that is transported along a transmission line moves in the electromagnetic field between conductors. The velocity of propagation is $c/\varepsilon_R^{1/2}$, where c is the velocity of light and ε_R is the relative dielectric constant. The only area along the transmission line that radiates or cross couples is where the wave is changing amplitude. For a step function, this area is the leading edge, as it moves down the line. As the rise time increases, the radiating area also increases. As an example, a wave travels 1.5 cm during a rise time of 100 ps. If the line is series terminated at the logic source and the line is 6 cm long, the radiation pattern lasts for one round trip of the wave or for 800 ps. The radiating pattern is the same for the outgoing and reflected wave. The rise-time frequency for a radiated signal is 3.2 GHz. In the case of a fast leading edge, the radiation at a distance from a transmission line appears to emanate from a moving current element, that is, a moving dipole.

3.7 NOTES ON RADIATION

Radiation and cross coupling occurs where there are changing field patterns. When the rise time is 1 ns, a wave can travel 20 cm during this time. For many circuit boards, logic lines are in transition over their entire length. In the section on cross talk, very short rise times are used to make it easier to describe the phenomena.

For outer traces (microstrip), most of the field is in the dielectric and a small part of the field is in air. The portion in air travels faster. This has the effect of smearing the leading edge and changing the nature of the cross talk.

When transmission lines (traces) are terminated, the energy carried on that line is dissipated in the termination. Cross coupling and radiation stops when wave action stops. For series (source) termination, the outgoing wave is 50% of the supply voltage. The radiation level is half the case where the line is parallel terminated. This radiation lasts for the round trip time of the wave.

For short transmission lines, terminating resistors may not be necessary. The energy stored on this line is then dissipated in the line and switch resistances. Again, the need for a terminating resistor is controlled by rise time and line length. Figure 3.5 shows the wave forms that take place on an unterminated transmission line with different rise times. This figure can be scaled. If the rise time is doubled the line line length is doubled.

The Rise-Time Rule. If the length of the transmission line is less than one-third the distance a wave travels in one rise time, then terminating the line in its characteristic impedance is not necessary. For a rise time of 100 ps, the rise-time distance is 1.5 cm.

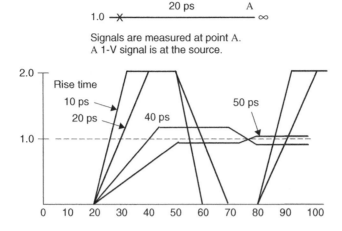

Figure 3.5 Rise time and the reflections on an unterminated transmission line.

One-third this distance is 0.5 cm. This means that terminations are needed for gigahertz clock rate logic. Series terminations are preferred. This same rule applies in the case of stub lengths.

3.8   THE CROSS COUPLING PROCESS (CROSS TALK)

The cross coupling process is complicated, because there are so many cases to consider.

Logic can be carried in either direction on any one line. Logic can be carried in the same or in opposite directions on parallel lines. Culprit lines can be longer or shorter than victim lines. Lines can be series or parallel terminated. Lines can be unterminated, which causes voltage doubling. Coupling on microstrip is different than coupling on stripline. Cross coupling depends on the logic family where there are different error margins.

The language of cross coupling has evolved with time. In radio transmission, terms such as *far-end* and *near-end cross coupling* are used. Interference acronyms such as FEXT for far-end cross coupling and NEXT for near-end cross coupling are used.

Cross coupling can be considered to have inductive and capacitive components with waves traveling in both directions. We will refer to forward coupling when the coupled signal travels in the same direction as the driving or culprit wave. Reverse coupling implies that the coupled signal travels in a direction opposite to the driving or culprit wave.

The cross coupling process between parallel transmission lines can be analyzed by assuming that the lumped parameter elements that make up a transmission line are loosely coupled. This assumption allows us to separate the coupling mechanism

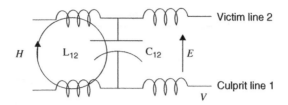

Figure 3.6 The mutual inductance and mutual capacitance coupling in parallel transmission lines. Note: Transmission lines are over a ground plane.

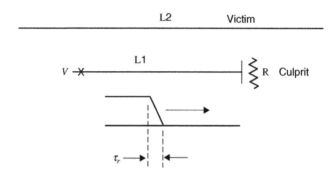

Figure 3.7 A step function wave applied to the culprit line.

into capacitive and inductive components. The representation in Figure 3.6 shows two traces over a ground plane. One of the lines is the victim and the other is the culprit.

When a logic voltage is switched on the culprit line L1, a wave propagates to the right. Note that in our preliminary discussion, the victim line L2 is longer than line L1. This allows us to discuss reverse coupling without having a nearby reflection point on the victim line. A culprit step function with a finite rise time is shown in Figure 3.7.

The coupling process involves both mutual inductance and mutual capacitance. We must equate the changing H field that is present in the inductance of L1 to a changing current and the changing E field that is present in the line capacitance to be a changing voltage. There is no loss in generality if we limit our discussion to a positive culprit wave that travels to the right. To start the discussion we will not consider terminations or reflections on the victim line.

3.9 MAGNETIC COMPONENT OF CROSS COUPLING

The magnetic flux generated by the culprit line current crosses the conducting path of the victim line. By Lenz's law, a changing current on the culprit line induces a negative current on the victim line. This current is associated with a negative voltage wave that travels to the right and a positive voltage wave that travels to the left. The associated fields have directions that satisfy the requirements of Poynting's vector.

In Figure 3.7, the culprit line is short and it couples to the middle of the victim line L2. Notice L1 is terminated in its characteristic impedance so the culprit wave is absorbed when the wave reaches the end of the line.

As the culprit wave progresses along L1 the only magnetic flux in transition is along the leading edge. In effect, the coupling is limited to the region where flux (current) is changing on L1. The coupling takes energy from L1 and adds it to L2 on a continuous basis. Each increment of transmission line couples an increment of current to a forward and reverse traveling wave. The coupled forward wave and the culprit wave move to the right at the same velocity, so there can be no coupled energy ahead of the leading edge. The induction process causes a negative pulse of voltage on L2 that increases in amplitude as the culprit wave moves to the right down the line. The pulse width is equal to the rise time of the culprit wave.

A reverse wave moves to the left on L2 as the culprit wave moves to the right. As the leading edge moves to the right, every increment of transmission line couples energy to a victim wave, as it moves to the left. In this mode of coupling the reverse wave reaches a fixed amplitude after the initial rise time. The wave that moves to the left is thus a step function of fixed amplitude. If the culprit wave reaches its termination at time t, the end of the reverse coupled wave will reach the starting point in time $2t$. The victim step wave lasts twice as long as the culprit wave. We will see that this reverse coupled wave will be the source of most coupling problems.

The amplitude of the reverse magnetically coupled step wave is given approximately by

$$A_{RL} = \frac{VL_M}{4L} \tag{3.2}$$

where V is the culprit voltage and L_M is the mutual inductance per unit length. L is the inductance per unit length used to calculate the characteristic impedance of the victim line. The amplitude of this reverse coupled wave does not depend on rise time or fall time. It only depends on the ratio of inductances. The reverse wave lasts twice as long as the forward wave. The rise time of the reverse wave is double the rise time of the culprit wave. Note that half of the coupled current is supplied to the forward wave. These two factors of two account for the factor of 4 in Equation 3.2.

The forward current pulse caused by inductive coupling is given by

$$I_{FL} = -\frac{1}{2}\left(\frac{L_M}{L}\right)\left(\frac{V_t}{\tau_r}\right)Z_0 \tag{3.3}$$

where t is the time of transit and τ_R is the rise time. The factor of 1/2 results because half the coupled current goes into the reverse wave.

The inductively coupled wave forms are shown in Figure 3.8.

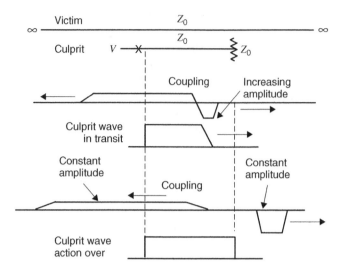

Figure 3.8 Inductive coupling between transmission lines.

3.10 CAPACITIVE COMPONENT OF CROSS COUPLING

Capacitive coupling contributes to waves traveling in both directions on the victim line. The coupled current that flows to the right is positive and the coupled current that goes to the left is negative. These polarities satisfy the requirements of Poynting's vector.

The capacitively coupled reverse voltage pulse is approximately given by

$$A_{RC} = \frac{VC_M}{4C} \tag{3.4}$$

where V is the culprit voltage, C_M is the mutual capacitance per unit length, and C is the capacitance per unit length used to calculate the characteristic impedance of the victim line. The reverse wave lasts twice as long as the forward wave and half the current is supplied to the forward wave. This accounts for the factor of 4 in Equation 3.4.

The rise time of the reverse wave is double the rise time of the culprit wave.

The amplitude of the forward capacitively coupled current component is given by

$$I_{FC} = \frac{1}{2} \left(\frac{C_M}{C} \right) \frac{Vt}{Z_0 \tau_r} \tag{3.5}$$

where V is the culprit voltage, t is the time of transit of the culprit wave, and τ_r is the rise time. The factor 1/2 results because half the current goes into the reverse wave.

The capacitively coupled wave forms are shown in Figure 3.9.

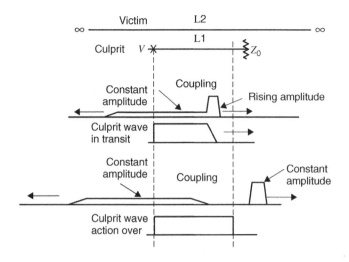

Figure 3.9 Capacitive coupling between transmission lines.

3.11 CROSS COUPLING CONTINUED

In a system of conductors, the propagation velocity depends only on the dielectric constant. A change in velocity does not change the rise time of an event. It only changes the distance a wave travels in a given time. In a given dielectric, the magnetic field and the electric field travel at the same velocity. The partitioning of fields that we have just used does not change this fact.

The voltage in the victim forward wave is the sum of the inductive and capacitive currents given by Equations 3.3 and 3.5 multiplied by the characteristic impedance or

$$A_{\text{SF}} = \frac{1}{2}\left(\frac{C_{\text{M}}}{C} - \frac{L_{\text{M}}}{L}\right)\frac{Vt}{\tau_{\text{r}}} \tag{3.6}$$

The ratio C_{M}/C is approximately equal to L_{M}/L. To a first approximation, the terms cancel. For traces stacked one above the other, the capacitive term may dominate. For side by side traces, the inductive term may dominate. Because A_{SF} is proportional to t and inverse to τ_{r}, this mode of cross talk can be a problem when trace lengths are long and rise and fall times are very short. On outer traces, (microstrip) the ratio of C_{M}/C is reduced because part of the E field is in air. This usually results in an increased inductive coupling. This would suggest that to avoid cross coupling, longer traces be run as stripline.

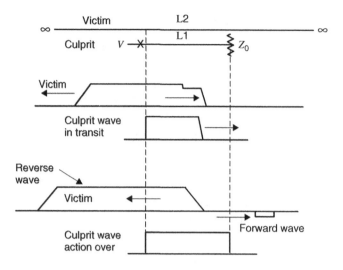

Figure 3.10 The total cross coupling between traces.

N.B.

Forward wave coupling increases as rise time decreases.

The reverse coupled wave is the sum of Equations 3.2 and 3.4 or

$$A_{SR} = V \left(\frac{L_M}{4L} + \frac{C_M}{4C} \right) \tag{3.7}$$

There is no cancellation of terms in this equation. This cross coupling wave is the one that usually creates the greatest threat to signal integrity. This cross coupling is shown in Figure 3.10.

3.12   CROSS COUPLING BETWEEN PARALLEL TRANSMISSION LINES OF EQUAL LENGTH

Parallel transmission lines can be series or parallel terminated with the same or opposite signal directions. Culprit and victim logic transitions can be in the same direction or in opposite direction. The culprit and victim can be unterminated at either end. There can be a culprit on both sides of one victim. A complete analysis of signal integrity requires an analysis of each case. In the following discussion, we will assume that the culprit wave is positive and goes to the right.

Consider the case where the culprit line is series terminated and the culprit makes a round trip during the rise time. The cross talk is composed of reverse coupling for half the round trip time and forward coupling during the return time. The coupled signal is maximum when the culprit signal has made a round trip. This line length is called the *critical length*.

The distance a wave travels in a rise time is called a *TEL* that stands for *transitional electrical length*. Lines longer than one-fourth TEL that are not terminated are apt to fail. In the example above, the critical length is half of TEL.

The following cases are a sample of the cross talk problem. A treatment of every case would be of little value, as there are just too many variables.

Case 1. Victim line is series terminated at the left.
At the point of first coupling, the reverse wave is absorbed in the victims matching impedance. There is no reflection. The forward pulse has a width equal to the rise time of the culprit and is usually of small amplitude. This pulse is reflected at the open end of the victim line, and it is not present at the next clock time. If the victim is at logic one and the coupling exceeds the clamping voltage then a reflection will result.

Case 2. Victim line is open circuited at the left.
The amplitude of the forward coupling on the victim line is proportional to time. When this pulse reaches the victim's series termination, the wave is absorbed and poses no problem. The reverse coupled wave is doubled and is reflected forward. If the doubled voltage exceeds the clamping voltage then a reflection or damage can result. This reverse wave lasts twice as long as the culprit wave, so it could add to the logic signal at the next clock time. Under some conditions this could cause a logic error.

Case 3. Forward crosstalk when the victim line is a zero impedance at the left.
In this case, the victim line is parallel terminated at the right end of the line. The victim voltage, when the coupling starts, can be a logic 1 or a logic 0. Logic errors are apt to occur when the coupled wave drives the logic state toward the transition region. In this region, the logic state is undefined. This case is shown in Figure 3.11.

N.B.

In any logic design, the error budget must allow for all modes of cross coupling.

The coupling between traces is often likened to transformer action where the transformer coils are single turns. This approach does not consider the character of the forward and reverse coupled waves.

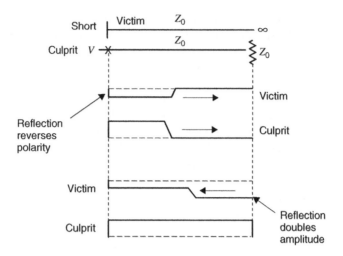

Figure 3.11 Coupling to a transmission line with a zero impedance source.

Guard traces are sometimes used in an attempt to reduce cross coupling. The approach assumes that added grounds act as shields. The approach is flawed because these traces might function as resonant couplers. The wide spectrum associated with logic transmission means that logic signals can cause ringing in any nearby unterminated transmission lines. The result is added coupling rather than shielding.

3.13 RADIATION FROM BOARD EDGES

The energy stored between conducting planes is in parallel with the energy stored in decoupling capacitors. When a demand for energy is made, waves propagate into both the decoupling capacitors and into the space between the planes. Reflections return energy to the point of demand. This wave action was discussed in Section 2.4. Consider waves that travel between ground and power planes. When a wave reaches the edge of the board, the open circuit reflects the wave and doubles the voltage. Wave fronts reach the board edge at different times. Waves double in amplitude as they reflect. The wave amplitude is inversely proportional to the distance from the point of demand. If the wave amplitude is 1 mV at the board edge and the planes have a 2 mm spacing, then the E field strength is 2 mV/2 mm or 1 V/m. This E field level would be measurable.

N.B.

Vias that are used to take energy from the ground/power planes should be located away from board edges.

In multilayer boards, each ground/power plane stores field energy. Energy is also stored in parallel decoupling capacitors. Every logic transition requires field energy. This energy comes from the nearest sources, which includes decoupling capacitors, ground/power plane capacitances, and connected logic. When logic crosses layers, it carries field energy to a new region. This energy must eventually be radiated or dissipated. Field energy cannot travel through conductors. It can only travel through holes in conductors (vias) or it can leave at the board edge. When logic lines are terminated, most of the energy used for driving the transmission lines is dissipated in the terminations. For unterminated lines, the field energy must be lost in trace and switching resistances and in radiation. The path that supplies the transmission line energy is the same path that is used to dissipate the energy (see the discussion in Section 4.16).

3.14   GROUND BOUNCE

Ground bounce often refers to the voltage measured between two ground points on a circuit board. It may also refer to the voltage drop on the ground pin of an IC. Since it is not possible to contact the die, the total voltage drop cannot be observed. Our point of view is to consider that energy supplied to the IC flows in fields on transmission lines. The short section from the board to the die is not a match to the impedance presented by the ground and power plane. The reflections on this short segment of line cause fields that would be sensed by a probe in this area. To attribute this field to the ground connection alone would be an error.

Measurements are usually made using an oscilloscope probe where the probe common is connected to one ground point on the circuit board and the probe tip is connected to a second ground point. The voltage that is sensed is interpreted as an IR drop in the ground plane. The measure is usually made when there is some sort of problem involving signal integrity and the source of difficulty is under consideration.

The presence of a probe in a circuit means that a short stub has been added. We have shown that even short stubs added to a transmission line can affect rise time. When the probe is used to measure points along a ground plane, the measurement has little effect on any logic transition.

Any measure of voltage is a measure of an electromagnetic field. Because the probe tip and the probe common (ground) form a loop, the probe will measure the field in this loop area. The field patterns in the area will probably be changed by the presence of the probe and its connections. In effect, the very field pattern that is sensed will be modified by the presence of the probe.

To begin a test, it is wise is to connect the probe tip to the probe common and make no connection to the circuit common. There should be no signal. Now, touch the probe common to the circuit common. Again, there should be no signal. Now, it is proper to move the probe tip to a second common point (ground) in the circuit.

The probe ground and probe tip form a loop. This loop can couple to any fields generated on the board. There is no simple way to orient this loop to avoid this coupling. Some of the signal may be the field associated with an IR drop, but usually, the largest component of the field will be associated with nearby logic transitions. At this point, it is important to note that the fields from transmission lines (traces and ground planes) are tightly controlled. The conductor geometry near component pins is not as well controlled, and this is often the source of field generation.

The real point to recognize is that there are IR losses in the ground plane. These losses are not simple to measure. Logic currents flow in a narrow paths under each circuit trace. The current flows where the E field lines terminate. For fast logic, in particular, the current does not penetrate very far into the conducting plane. The limited amount of copper that is used means that there is resistance that must be considered (see Section 4.11 for a discussion on transmission line resistances).

3.15 SUSCEPTIBILITY

There are often applications where logic structures must be placed in hostile electrical environments. A hostile environment might be the proximity to high current switching. ESD is certainly a big concern for many applications. In large systems, the issue of lightning protection must be considered. The solution to protection will vary depending on whether the logic is to operate without error or simply survive the interference.

Interference coupling is always field coupling. Different interference types have energy associated with different parts of the frequency spectrum. For this reason, different approaches are needed for each type of interference. A lightning pulse is associated with a 100,000 A with rise-time frequency of 640 kHz. An ESD pulse is 5 A with a rise time frequency of 300 MHz.

The control of interference coupling is almost always associated with loop area control. This is the same control used in designing logic structures. The loop areas associated with on-board logic transmission are small. This means that the transmission lines will not usually couple to external fields. The problem of greatest concern is coupling to cables that interconnect parts of logic systems. The subject of interference control is treated in the author's book "Grounding and Shielding and Interference Control," listed in the bibliography at the end of this book.

GLOSSARY

Common mode (Section 3.5): Refers to an average interfering signal that appears between all conductors in a cable and the local ground. The term finds use in working with balanced signal lines. It is often the ground potential difference between two pieces of hardware. A common-mode signal is not always interference. There can be many common-mode signals present at the same time. In logic, common mode is called *even mode*.

Cross coupling (Sections 3.4 and 3.8): The unwanted coupling of signals between traces on a circuit board.

Culprit (Section 3.8): The signal causing coupling or interference.

Critical length (Section 3.12): Half the distance traveled by a wave in its rise time.

Dipole (Section 3.2): A conductor geometry used as an antenna. Two conductors in the shape of outstretched arms driven at the midpoint by a balanced signal.

Fall time (Section 3.4): The time for a step function to change from 80% to 20% of final value.

Far field (Section 3.4): The electromagnetic field far from a radiating source.

Forward coupling (Section 3.8 and 3.11): The cross coupling that moves in the same direction as the culprit wave. Forward coupling is a pulse that lasts as long as the culprit rise time. Forward coupling increases with time and is most severe on long lines.

Ground bounce (Section 3.14): The IR drop along the ground plane or on conductors between a ground pad and the die.

Induction field (Section 3.4): The field near a loop where the H field dominates; a low impedance field; a magnetic field.

Impedance of free space (Section 3.4): The ratio of E- to H field intensity in the far field or 377 ohm.

Interference (Section 3.1): Any unwanted signal that couples to a circuit. All coupling is field related. Coupling associated with current flow is called *conductive coupling*. Sometimes called *common-impedance coupling*.

Near field (Section 3.4): The electromagnetic field near a radiating source.

Near-field/far-field interface distance (Section 3.4): The distance from a radiator where the field can be considered a far field. Beyond this distance, the wave impedance is 377 ohm. These waves are called *plane waves*.

Noise budget (Section 3.1): In logic, the signal level fluctuations that are permitted for normal logic operations.

Normal mode (Section 3.5): The normal signal. In balanced logic, the signal is called *odd mode*. In analog systems, it is the signal of interest.

Plane waves (Section 3.4): Electromagnetic waves beyond the near-field/far-field interface distance. This definition is applied to sinusoidal radiation.

Radiation (Section 3.3): The field energy that leaves a circuit and does not return.

Reverse coupling (Section 3.8): The wave that is cross coupled that moves in the direction opposite to the culprit wave.

Rise time (Section 3.4): The time for a wave to change from 20% to 80% of final value.

Rise-time frequency (Section 3.4): The frequency determined by rise or fall time in the equation $1/\pi\tau_r$. A signal at this frequency can be used in an analysis to determine the magnitude of response to a transient input.

Transitional electrical length or TEL (Section 3.12): The distance a wave travels in one rise time.

Victim (Section 3.8): The trace influenced by cross coupling or interference.

Wave: A general term implying the movement of electromagnetic energy.

Wave impedance (Section 3.4): The ratio of E field intensity to H field intensity at a point in space. This ratio is the characteristic impedance of a transmission line. Near a radiating source the ratio changes depending on the nature of the radiator. In the far file the ratio is 377 ohm.

ENERGY MANAGEMENT

4.1 INTRODUCTION

It is interesting to consider the way energy is supplied to a switched electrical load in a building. For our discussion, pick a moment in time when the voltage between the "hot" lead and the grounded power conductor is at its peak of 170 V. Note that this peak of voltage appears at all points in the building at the same time.

The events we will consider will last no longer than 1 ms. At $t = 0$, a switch closes placing a 17-ohm resistive load across the line. Assume that the power wiring looks like 100-ohm transmission lines. At the moment of switch closure, the voltage drops to 24.7 V. This means that a step wave of -145 V starts to travel down the power wiring. When this wave reaches a branch point in the wiring, the wave divides. A part of the wave reflects and travels back to the load. At the load this wave divides again and a fraction of the wave adds voltage to the load. The reflected part of this wave then travels back to the first branch point. A part of the first wave that reached the first branch point continued outward toward other branch points. In a short period of time, waves will have traveled and reflected all over the facility. Each time one of these reflected waves returns to the load the voltage is increased.

Digital Circuit Boards: Mach 1 GHz, First Edition. Ralph Morrison.
© 2012 John Wiley & Sons, Inc. Published 2012 by John Wiley & Sons, Inc.

As the waves travel and reflect, the voltage at each branch point first drops and then begins to recover. What is happening is the energy stored in the electric field in the building wiring is being moved around to supply energy to the load and to even out the voltage in the facility. In the first microsecond, waves have traveled throughout the building, and there have been hundreds of transmissions and reflections. Some of the waves will have actually reached the service entrance and will have moved out onto the distribution wiring. If the power wiring is in conduit, very little of this wave energy will radiate.

A fast oscilloscope placed at the load will show the initial voltage sag. Within the first few microseconds the voltage at the load will recover to near 170 V. If the voltage is viewed at nearby branch points, the voltage will sag but not to the same degree. If the line voltage is observed at the service entrance, the first sag may only be a few volts. Beyond the service entrance, the sag in voltage is hardly noticeable. The transmission lines in the building are not crafted for high frequency performance, so there are all sorts of discontinuities. Very soon, the waves are smeared, and there is no way to identify them as specific transmissions or reflections.

The energy demanded by the load must eventually come from a power generator. The only way a generator can sense that it must put out more power is for there to be a sag in voltage at the generator. This sag is made up of a large number of wave reflections that travel from the building to the generator. The request is a slow drop in voltage and not a sharp spike as measured at the switch contact.

It takes a few milliseconds before a voltage demand reaches the generator. Several seconds later, the generator can respond and raise the voltage. Even if this response is carried as waves to the load, the wave action cannot be identified.

After the switch closed, the energy was supplied to the load from the local electric field. The energy stored in the electric field in a building can be easily calculated. If there is 600 m of wiring and the capacitance per meter is 333 pF, the total capacitance is 0.2 μF. The energy stored in the capacitance at 170 V is 5.4 mJ. If this energy were supplied in 1 μs, the power level would be 5400 W. The peak power demanded by the load in this example is 1700 W. This calculation shows that the electric field energy that is stored in the building is capable of supplying the initial energy to the load.

There are many parallels between the switching action on a circuit board and the switching action in a building. Assume that a fast logic switch closes and connects a logic trace to the power supply. At the moment of switch closure, the voltage will sag based on the characteristic impedances of the immediate connected traces. Figure 4.1 shows traces that might be involved.

The connections include logic traces, traces to a decoupling capacitor and vias that connect to the ground/power plane. Those logic traces that are already connected to the power supply voltage store electric field energy in their transmission lines. The voltage sag is based on the characteristic impedances of these parallel lines.[1] Waves

---

[1] The logic traces that are connected to the power leads inside an IC must sag when the power voltage sags. All of these connected logic conductors will then supply some energy to any new loads switched on to the IC. Reflections on these logic lines will return energy to the switched load.

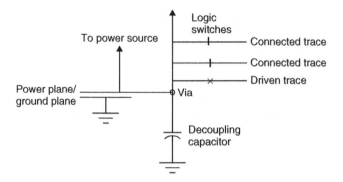

Figure 4.1 The traces involved when a logic switch closes. Note: All traces and connections are transmission lines. This is not obvious in a schematic representation.

with a negative voltage will travel out on all the power connections at the IC to the decoupling capacitor and to the ground/power plane. The reduced voltage will also progress forward on the newly connected logic. When these outgoing waves reflect and return to the load, they bring back energy that increases the voltage at the IC. Many reflections must take place before the voltage rises to near its final value. In a facility, a voltage sag lasting 1 μs is acceptable. In a logic circuit, a voltage sag of 30% lasting nanoseconds can cause trouble. It is this problem that we address in this chapter.

When the power supply sags, the traces that carry logic signals to the IC can be at a voltage higher than the power supply. If protecting diodes become conductive, then some of this voltage sag will be transmitted on these traces back to logic sources. If these logic lines are series terminated, this wave energy will be absorbed in these resistors. The point to be made is that the power supply sags can propagate on both the incoming and outgoing logic lines and must be considered a part of the noise budget.

4.2   THE POWER TIME CONSTANT

Circuit theory supports the view of instantaneous power. When a 50-ohm load is placed across a 5-V voltage source, we expect that the power level to be 0.5 W immediately. What actually happens in the first few nanoseconds is different. Assume a fast switch and an ideal voltage source in series with a 50-ohm transmission line. When a 50-ohm load is placed on the far end of the transmission line, the voltage sags to 2.5 V. The power level at this time is 0.125 W. The power will stay at this level until a wave moving on the line reflects at a voltage source and returns to the load. Waves must make several round trips before the power level rises to near 0.5 W. If the line is 1 cm long, one round trip time on a typical circuit board is 0.13 ns.

Consider the case when the 50-ohm transmission line is connected to a 5-V source and the load is 5 ohms. The expected power level is 5 W. On a 50-ohm line the voltage sags to 0.45 V and the delivered power is only 0.0409 W. The sag in voltage means that a negative wave of 4.55 V is sent down the line toward the voltage source. A reflected wave at the voltage source brings energy back to the load. Waves continue to make round trips between the load and the voltage source until the voltage at the load rises to near 5 V. The time constant as given by Equation 2.7 depends on the parameters of the transmission line and the value of the load. For a 5-ohm load and a 50-ohm line that is 10 cm long, the time constant is 14.3 ns. It takes about 3 time constants to reach 95% of final value or about 44 ns. For many logic circuits this delay is not acceptable. See Figures 2.12 and 2.13 for examples of how the voltage builds on cascaded transmission lines.

N.B.

On any transmission line it takes time for energy flow to change to a new level.

N.B.

The power level at the moment of demand depends on the characteristic impedance of the source.

In circuit analysis we are familiar with the term *ideal voltage source*. When a load is applied to an ideal voltage source the current that flows is instantaneous. The voltage must remain constant regardless of the load value or how long the load remains connected.

If this ideal voltage source existed, any connected load involves some lead length. In fact, every component has dimensions, and this means there is always a transmission line process involved when a load is connected to a voltage source. A 10-mm-long connection is small, but this connection is still a transmission line. Again using Equation 2.7, the time constant is 4.4 ns. At 1 GHz a clock period is 1 ns, thus it is obvious that short connections limit power flow in the first few nanoseconds. This delay must be considered in the design of fast logic.

In theory, this connection problem can be solved by reducing the characteristic impedance of the leads connecting to the source. For example, if the characteristic impedance of the 10-mm line is 0.1 ohm, the voltage sag would only be 2%. This assumes that the load has no lead length, which is of course impossible.

N.B.

When a load is connected to an ideal voltage source, the rise time depends on the load impedance, the lead length, and the characteristic impedance of the connection.

The loop area involved in connecting a load to a voltage source certainly has inductance. In a circuit sense, this connecting lead geometry plus the component capacitance form an RLC circuit. We have shown that in a fast circuit, any time delay that is measured involves the round trip propagation time and multiple reflections through the entire circuit. This means that in fast circuits, smooth exponential transitions do not occur. For example, during the time when the first wave is making a round trip, the voltage is constant.

4.3 CAPACITORS

The classic capacitor consists of parallel conducting planes separated by a dielectric. The symbol for a capacitor implies that connections are made to a mid point on the planes. Many practical capacitors are made by stacking several conducting planes and making appropriate connections to the conductors. To be effective, a capacitor should supply current on demand with a limited sag in voltage. For a step function load, this requires a low characteristic impedance at the capacitor terminals.

N.B.

All two-terminal capacitors are short transmission lines. This line is connected to pads, which is a second transmission line.

To simplify the discussion, consider a two-terminal capacitor as shown in Figure 4.2.

For a two-terminal capacitor, energy must enter and leave through the same port. When a load is applied to this geometry, energy is supplied to the load from the capacitor and from any connecting transmission lines (traces). This includes logic lines connected in the IC. The majority of the energy eventually comes from the capacitor. After the load is connected, waves travel into the capacitor and onto any other connecting transmission lines. The wave that travels into the capacitor, reflects

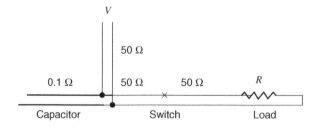

Figure 4.2 A capacitor as a transmission line in a circuit. The values shown are characteristic impedances.

at the open end and brings back energy to the load.[2] The wave(s) that follows the connecting trace(s) reflect at branch points also bring back energy to the load and to the capacitor. One of the paths can connect to the ground/power plane capacitance. This complex of wave actions replenishes the charge in the capacitor.

Any resupply of energy to the capacitor must come from multiple reflections on traces with a nominal characteristic impedance of 50 ohm. Even if the capacitor is parallel connected to the ground/power plane capacitance, this energy must flow through a connection (vias) that also have a characteristic impedance of approximately 50 ohm. As we have seen, waves must make multiple round trips through the vias to supply the required energy. Obtaining energy has a time constant that depends on transmission line lengths and characteristic impedances.

N.B.

The supply of energy to a two-terminal capacitor is restricted by the fact that the load reduces the voltage at the supply point. At the moment of demand, the energy can only flow out of the capacitor.

The analogy with a water reservoir is appropriate. The water that is supplied to a reservoir does not use the exit conduit. This water always enters through a separate port. Even though a conduit is a two-way device it is futile to try to force water to flow in both directions at the same time. This is exactly what we attempt to do when we draw energy from a capacitor, and at the same time try to put it back through the same two terminals.

Capacitors are often constructed using stacked conducting layers. This design approach is an effective way to increase the capacitance in a small volume. The problem with this construction is that the inner layers can be blocked from receiving energy by the parallel connections at the ends of the capacitor. The energy that enters or leaves these layers of the capacitor must use the air space around the sides of the capacitor. In a circuit sense, this has the effect of raising the entry inductance.

4.4 THE FOUR-TERMINAL CAPACITOR OR DTL

The most effective way to provide fast decoupling is to build the capacitor as a four-terminal transmission line. Energy is supplied from one end and taken from the

---

[2]The idea of "bringing back energy" is explained as follows: When a load is first connected to a transmission line, a wave carries energy into the load from the electric field in the transmission line. The wave that leaves the point of connection reflects from every discontinuity. When this discontinuity is a branch circuit, the reflected wave usually carries energy back to the load. When this wave reaches the load, another reflection occurs. A fraction of this wave propagates (transmits) into the load increasing the voltage. The capacitor is the dominant transmission line of the system. In a sense the reflection of waves brings energy back to the capacitor and the load.

other. In effect, this is a transmission line positioned between a load and a source of energy. We can call this a decoupling transmission line or DTL. The source impedance of this DTL determines the voltage drop when a demand is made for current. The length of the DTL determines how long this voltage is sustained. After a demand for current, the voltage drop depends on the ratio of characteristic impedances. A small wave makes a round trip in the DTL. When the wave returns to the load, the voltage will drop a second time. A loaded four-terminal capacitor is shown in Figure 4.3.

A practical DTL could have a source impedance as low as 0.1 ohm. A 1-A demand at the load end would cause a voltage drop of 0.1 V. When the resulting wave reaches the far end of the DTL, a wave is both reflected and transmitted depending on the parameters at the discontinuity. If the far end is a short transmission line connected to a voltage source then many round trips can take place in this short line before a second wave makes a round trip in the DTL. Section 2.15 discussed the flow of energy between mismatched transmission lines. The balance of energy flow is given by Equation 2.17. If this condition is met, there will be very little sag in the voltage supplied by the DTL.

The capacitance of a DTL can be expressed in terms of its characteristic impedance and the time it takes for a wave to travel the length of the DTL. If the characteristic impedance is Z and the travel time is t then the capacitance is given by

$$C = \frac{t}{Z} \tag{4.1}$$

As an example, if $Z = 0.2$ ohm and $t = 1000$ ps, the value of C is 5000 pF. If the voltage is 5 V and the load is 5 ohm, the voltage will drop to 4.8 V and remain constant for $2t$ or 2000 ps. Because this voltage is constant, it is correct to say that the source has no series inductance. In practice, any load that is connected to a DTL is another transmission line.

It is interesting to consider the power distribution structure on a circuit board as a network of transmission lines. Every decoupling capacitor that is placed between

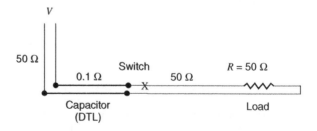

Figure 4.3 A four-terminal capacitor (DTL) in a circuit. The values shown are characteristic impedances.

power and ground is in effect with two-transmission lines in series. The first transmission line is the capacitor and its dielectric that stores a charge and the second line is the connection to the capacitor. These two transmission lines are in effect two-short stubs in series.

N.B.

The source impedance of a DTL could be 50 times lower than that of 10 parallel capacitors.

When a DTL is placed between a power source and a load, it can provide a 0.1-ohm link. In effect, it looks like a power filter. The DTL reflects voltage fluctuations that arrive from either direction. This, in effect, splits the power structure into two sections. If a DTL is placed next to a noisy source, only a small fraction of the noise will propagate to the other side of the DTL. For this reason, the DTL is a powerful tool in controlling both conducted and radiated interference. It provides decoupling in the full sense of the word.

N.B.

The lead length at the load end of a DTL should be kept short, if voltage swings are to be limited.

4.5 TYPES OF DTLs

There are several transmission line geometries that can be used in constructing a DTL.

The simplest geometry is a pair of flat conductors separated by a dielectric. A voltage source is connected to one end and the load is connected to the other.

N.B.

The connections to a die are usually through bonding leads that are a section of 50-ohm line.

A second approach is to form the DTL as a coaxial cable. As an example, small round conductor can be coated with a dielectric. A second conductor is then plated over the dielectric. Several of these coaxial filaments could be routed in parallel forming a DTL. Using this technique, the characteristic impedance in theory could be milliohms.

Another approach is to build the DTL transmission line as two closely spaced traces. The traces could be folded in the form of a maze or carried over some distance. A maze configuration could also be a part of a die.

4.6   CIRCUIT BOARD RESONANCES

In passive circuit theory, an inductor and a capacitor in series or in parallel are resonant circuits. To illustrate the parallel resonance idea, consider a capacitor with stored energy. At time $t = 0$, the capacitor and inductor are connected parallel. Initially the current in the inductor is zero. Current will start building in the inductor. At some point all the energy stored in the capacitor is transferred to the inductor. At this time, the voltage across the capacitor is zero and the current in the inductor is at a maximum. The cycle continues with the charge building in the capacitor until the voltage is a negative maximum. If there are no losses, this back and forth movement of energy will continue indefinitely. The voltage across the circuit is a sine wave. At the peaks of voltage, all the energy is stored in the electric field of the capacitor. At the zeroes of voltage, all the energy is stored in the magnetic field of the inductor. The key factors in this resonance are the transfer of energy between the electric and magnetic fields and the time it takes for the energy to make the transfer.

A section of unterminated transmission line can be viewed as a type of resonant circuit. When a voltage V is switched on the line, a wave propagates down the line. At the open end of the line, a reflection takes place and a wave returns to the source. The reflection at the source returns this wave and the voltage source supplies no more energy to the line. Wave action continues indefinitely with the voltage at the open end alternating between 2 and 0 V. The total energy on the line is constant with the energy in constant motion. Just as before, the moving energy is stored in both the magnetic and the electric fields. The time of one cycle is twice the propagation time of the wave on the line. This is a step function resonance rather than a sinusoidal resonance. In parts of the transmission line where the voltage is zero, energy is stored in the inductance; and in other parts of the line where the current is zero, the energy is stored in the capacitance. The exchange of energy takes place at the leading edge of the wave.

When an unterminated stub is added to a terminated transmission line, the energy flow in the stub acts very much like a resonant circuit. If the stub is long enough, the energy moving back and forth in the stub can delay the settling time on the main line.

N.B.

A circuit board is a maze of transmission lines. If the lines are unterminated then these resonances are excited by changes to the logic states. In a good design, the logic levels on a circuit board settle to near their final value within one round trip on the longest line.

4.7   DECOUPLING CAPACITORS

Capacitors that are used for power decoupling in digital circuits are usually surface-mount types. This means that there are no connecting leads. The leads that must be considered are connections to pads and connecting traces. Capacitance values vary from a few hundred picofarads to 10,000 pF. Most of the capacitors that are used for decoupling use a ceramic dielectric. Manufacturers will usually provide information on the equivalent series resistance (*ESR*) and the equivalent series inductance (*ESL*), as measured at the terminals.

For high frequency sinusoids, capacitors are usually considered series resonant circuits. At their natural resonant frequency, the series impedance is a resistance. The series resistance at the resonant frequency is largely controlled by skin effect. A typical resistance is in the range 0.1–1.0 ohm. For a capacitor of 0.001 μF and a series inductance of 1 nH, the resonant frequency is 159 MHz. This is well below the frequencies involved in gigahertz operations. At frequencies above the resonant frequency of a capacitor, the impedance appears to be inductive. All of these ideas are related to sine wave analysis. In fast digital applications the transmission line view is preferred. This is the view point taken in this book. This view makes it easier to consider the response of a capacitor to step changes in load.

In designs involving high clock rates, it is customary to place many decoupling capacitors on the circuit board. To optimize performance, these capacitors are usually located close to circuits that switch the most energy. Some ICs require a cluster of capacitors. In effect, these parallel capacitors are groups of transmission lines having the same length. If the connecting traces are short, the circuit looks like one transmission line with a low characteristic impedance. This view becomes flawed when the capacitors are interconnected by a grid of transmission lines.

At 100 MHz, a clock period is 10 ns. A wave can travel in an epoxy dielectric about 150 cm in this time. This means that waves can propagate and reflect over an entire circuit board before the next clock time. This seems to imply that decoupling capacitors can supply energy to a load within one clock period. Taking energy from a decoupling capacitor requires waves to make many round trips on the connecting traces and internal to the capacitor. The resulting settling process can take more than 10 ns depending on the nature of the load and the location of the capacitor. If the power supply voltage swings, the wave action that results can cause radiation, and it can cross couple interference to nearby signal traces.

For clock rates of 1 GHz, the decoupling problems are significant. In one clock period, a wave in stripline can only travel about 15 cm.[3] For this logic to function, the first energy must come from connecting traces and from the ground/power plane capacitance. As we have stressed, all of these connections are transmission lines and multiple reflections are required to move energy over these lines.

The ground/power plane capacitance was discussed in Section 2.13. This is a tapered transmission line with a distributed characteristic impedance. It takes time to obtain energy taken from this conductor geometry. Most of the time is spent in the reflections through the via at the point of connection. Fortunately, the ground/power plane is a four-terminal source of energy.

---

[3] This assumes that the waves travel in a dielectric with a relative dielectric constant of 4.

When a capacitor is placed across a transmission line, it is equivalent to placing an open-ended charged stub across that line. Even if the capacitor has a characteristic impedance of 0.1 ohm, the connecting leads are a short section of 50-ohm line. The reflections depend on location and the geometry of the capacitors. If the capacitor is multilayered, it may look like a transmission line with an added small series inductance (Section 4.6). The problem we discussed earlier still exists. The stub can provide energy but it should be remembered that it only has two terminals. Trying to move energy in and out of a two-terminal capacitor at the same time is not possible.

4.8 THE BOARD DECOUPLING PROBLEM

A printed circuit board design usually includes a group of decoupling capacitors. These capacitors are connected across the power supply leads near active devices that drive transmission lines. They are usually parallel connected to ground/power planes using vias. It is often assumed that the planar capacitance supplies the first energy when there is a step demand. Because of the typical via geometry, this assumption may not be true.

Designers often consider each decoupling capacitor used on a circuit board as having a series inductance. In this analysis, all transmission line effects are ignored. Capacitors are staggered in value so that the natural frequencies associated with these series resonant circuits are spread out over the spectrum of interest.

In a circuit sense, capacitors are characterized by their natural frequencies. At this frequency, a capacitor resonates with its series self-inductance or *ESL*. At this resonant frequency, a capacitor looks like a resistance, which is called the *ESR*. Above this frequency, a capacitor looks like an inductor. The impedance of a capacitor including these parasitic is shown in Figure 4.4.

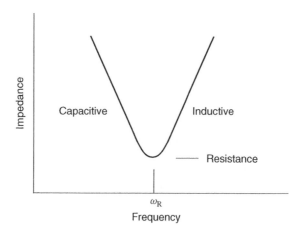

Figure 4.4 The impedance of a capacitor near its resonant frequency.

The resonant frequency of a capacitor depends on its construction style and capacitance. In general, a larger capacitor value implies a lower natural frequency. When different capacitor values are paralleled, the resulting impedance is that of paralleled series resonant circuits with different natural frequencies. Consider the frequency where capacitor 1 has an inductive reactance equal to the capacitive reactance of capacitor 2. At this frequency the parallel circuit looks like a parallel resonance, and the impedance rises. When different capacitors are parallel connected, the result is a complex impedance that rises and falls with frequency depending on how the resonances are distributed. When selected capacitors are paralleled, it is possible to limit the impedance to a maximum value over a range of frequencies. A design of this type can be handled on a circuit analyzer. The number of capacitors of any one value and style will vary with the smaller capacitors dominating the count.

N.B.

A circuit approach used to analyze the source impedance of a power supply ignores the delays associated with transmission lines.

A group of parallel decoupling capacitors can be used to supply energy to the active elements of a printed circuit board. For clock rates below 100 MHz this approach can be useful. Above 100 MHz, when transmission line delays are considered, the voltage source is no longer a set of parallel tank circuits. The performance relates to where the capacitors are located. For some devices, it has become necessary to locate a low inductance capacitor array (LICA) directly under the component in question. In this way, the connecting transmission lines are kept short. It should be remembered that there is conducted and radiated emissions out to 1 GHz even when the clock rates are 50 MHz.

4.9   THE IC DECOUPLING PROBLEM

IC dies are extremely dense component packages. Trace lengths are very short so that reflections in the die are normally not an issue. Consider the case where an IC must drive an external 50-ohm transmission line. The energy to drive the line would normally come from decoupling capacitors and the ground/power plane capacitance, both of which are external to the IC package. Assume that the path between the die and an external decoupling capacitor is a short section of 50-ohm line. When the logic switch closes, the voltage to the IC and to the external transmission line must drop. If the rise time of the switch is long enough (the clock rate is low enough), the voltage will not sag. For rise times less than 1 ns, there needs to be local decoupling on the die. Without this decoupling, the reduced voltage after a logic switch closure can cause the logic to malfunction.

If the die must drive a 16-bit parallel port with a short rise time, the current demand at switch closure can exceed 1 A. Assume that these lines are series terminated. This current must flow until the waves on the transmission lines have made one round trip. The energy demand is directly proportional to the length of the transmitting path. This level of energy cannot be supplied through a short section of 50-ohm line even if it is only 5 mm long. Supplying this energy from a decoupling capacitor located on or at the die is one solution. Connecting the die to an external decoupling capacitor using a transmission path with very low characteristic impedance is another possibility. This is the DTL solution described earlier. The length of this line is discussed in Section 2.15.

Dies with "bumps" can be mounted directly to a circuit board. This approach eliminates the transmission lines required on an interposer board. The problem of connecting the die to a low impedance source of energy on the board still remains. The dimensional ratios required to make a low impedance connection to a voltage source would seem impractical.

4.10   COMMENTS ON ENERGY MANAGEMENT

Circuit traces can literally support amperes of current flow, but it takes time for the current to reach these levels. To supply high step currents in nanoseconds, the characteristic impedance of the transmission path must be very low. As an example, if the characteristic impedance is 0.1 ohm and the voltage is 5 V, a load of 5 ohm will draw a current of 0.97 A with essentially zero rise time.

In an integrated circuit, many output logic traces can be connected to the source of voltage at any one time. If 10 lines are connected then the characteristic impedance at the logic junction is about 5 ohm. If an 11th logic trace is connected at clock time, the initial energy supplied to the connected line comes from 10 other lines in parallel plus the trace connected to power. In the other extreme, if no logic is connected and a logic trace is connected to power and if the source impedance is 50 ohm, the voltage will drop to 50%. This assumes that no energy storage is on the die. This example shows that the voltage levels at the die will vary depending on the number of traces that are connected and the number of traces being connected.

Consider a logic system where the number of drivers that demand current at any one clock time is fixed. Under these conditions the current demand will also be a constant. This arrangement does not require decoupling capacitors as there are no step demands for current. This approach would require dummy transmission lines to handle the case when all zeroes must be transmitted. The penalty for constant current flow is power consumption.

A complex IC may have over 10 million logic switches. On average, the current drawn by the internal logic is constant. Variations in current level of a few milliamperes can be handled by on-die decoupling. When line drivers are involved, the decoupling problem discussed earlier must be solved. If the drivers have a limited rise time then there is time to power the connected transmission lines from local energy storage.

4.11   SKIN EFFECT

Skin effect is usually discussed in terms of sine waves. When a conductor carries a sinusoidal current, the current is associated with a changing magnetic field. This changing magnetic field generates an emf that opposes the very current creating the field. The result is that a changing current tends to flow near the surface of a conductor. Because the inner copper has limited current flow, the conductor resistance appears to rise as a function of frequency. This rise in resistance with frequency is in addition to the reactance term that must be considered for an isolated conductor.

The equations for skin effect are often derived by applying Maxwell's equations to sinusoidal plane waves reflecting off of an infinite conducting plane. The depth where the field is attenuated by the factor $1/e$ is called a *skin depth*. This depth d is given by

$$d = \frac{1}{(\pi \mu \sigma f)^{1/2}} \qquad (4.2)$$

where μ is the permeability of free space, σ is conductivity of copper, and f is the frequency in hertz. For copper at 1 MHz, the skin depth is 0.066 mm. At this depth, the field strength is reduced to 37% of the value at the surface. At double this depth, the field strength is reduced another 37% to 13% of its surface value. Skin depth is inversely proportional to the square root of frequency. At 100 MHz one skin depth in copper is 0.0066 mm. There is field at the center of the conductor, but it is attenuated. Field strength and the current density are proportional to each other.

Equation 4.2 is used to approximate the skin effect for traces and round wires. When step functions are involved the waves are a composite of many sine waves. As a result Equation 4.2 serves only as a guide. Skin effect forces the higher frequency components of a step wave to flow nearer the surface. This has the effect of adding rise time to the step function, as it propagates down a line.

The resistivity of copper is 1.732×10^{-6} ohm cm. A trace has a dc resistance of

$$R = 1.732 \times 10^{-6} \frac{\text{length}}{\text{width}} \times \text{thickness} \qquad (4.3)$$

where the dimensions are in centimeters and R is in ohms. As the frequency increases, the resistance increases because of skin depth. Above 10 MHz the thickness of a practical trace does not enter into the equation and the resistance becomes

$$R = 8.24 \times 10^{-3} f^{1/2} \frac{\text{length}}{\text{width}} \qquad (4.4)$$

where R is in ohms and f is in gigahertz.

N.B.

The clock rate can be used as the frequency in calculating skin effect.

Trace thickness on a circuit board is usually under 2 mil. For fast logic, most of the current flow is on the trace surface next to the nearby conducting plane. A trace 5-mil wide and 2-mil thick has a dc resistance of 0.135 ohm/in. Skin effect raises this resistance at 1 GHz to about 1.2 ohm/in. The return current path on the ground plane is wider than the circuit trace. Since the skin depth is the same, the resistance of this path is about 0.8 ohm/in. The total series resistance of this transmission line at 1 GHz is therefore about 2.0 ohm/in.

A distributed resistance in a transmission line increases the characteristic impedance of the line. The increased impedance is given by

$$Z = Z_0 \left(1 + \frac{j\omega L}{R}\right)^{1/2} \tag{4.5}$$

If R is 1 ohm/in and L is 1 nH/in, the effect is less than 1% at 1 GHz.

The signal loss is given by the IR drop. On a 50-ohm line, a 3-ohm resistance causes a signal drop of 6%. This must be considered in an error budget analysis.

In electrostatics, it is shown that the presence of a voltage between conductors implies that there are surface charges. Consider taking energy from a charged transmission line. When a switch closes and a wave starts to move on this transmission line, it is these surface charges that move first. At the leading edge of the wave there is very little current penetration into the conductors. If the current pattern across the conductor could be observed at a fixed point near the point of demand, the current would appear to move into the center of the trace as a function of time.

On a series terminated transmission line the current flows on the line until the reflected wave cancels the current. The voltage along the line is tapered because of the distributed voltage drop. This tapered voltage drop is a distributed signal that propagates along the entire line. If there is a reflection at the open end of the line, this tapering continues until the wave stops. A tapered voltage is a signal that eventually adds to the voltage at the termination.

For traces over a ground plane, skin effect limits current flow to one side of the plane. The skin depth at 100 MHz is only 0.25 mil. If the conducting plane has a thickness of 2 mil, there are eight skin depths. The attenuation for eight skin depths is over 140 dB. This means that fields using transmission lines running parallel on opposite sides of a conducting plane are not coupled. This means that there is no significant cross talk through a conducting plane.

N.B.

There are several ways where energy can cross over the other side of a conducting plane. The crossing can be at a board edge, through a hole (via) or along a break in the conducting plane.

4.12 DIELECTRIC LOSSES

Dielectric losses are the result of molecular activity in the board. This dissipation is limited to the areas where a wave is in transition. This dissipation increases the rise time of a wave, as it progresses along a transmission line. The attenuation factor α in decibels per inch for sine waves is

$$\alpha = 2.3 \, f \, \tan(\delta) \varepsilon^{1/2} \tag{4.6}$$

where f is frequency in gigahertz and δ is the shift in phase resulting from transmission losses at the frequency f. The term *loss tangent* is applied to the factor $\tan(\delta)$. This factor is given in the literature for different materials. The value of $\tan(\delta)$ for FR-4 board material at 1 GHz is 0.02. For higher grades of dielectric, the loss tangent can be as low as 0.002.

Skin effect has a moderate impact on rise time. Note that skin effect causes the resistance of the line to increase proportional to the square root of frequency. Dielectric losses increase exponentially with frequency. This means that above some frequency, the dielectric losses will dominate. To see the effect on square waves, the amplitude of the fundamental might be attenuated by 0.1 dB/in. At the seventh harmonic, dielectric losses would dominate and the attenuation would be 0.7 dB/in. A 1-dB attenuation is a loss of about 10%. If the loss tangent is 0.02, the effect is to increase the rise time 8 ps/in of line. For a material with a loss tangent of 0.002, the impact is only 0.8 ps/in. Clock signals are apt to be affected by dielectric losses as the runs are usually longer.

4.13 SPLIT GROUND/POWER PLANES

Circuit board layout involves interconnection of many components. These interconnecting traces must have a specified width and spacing to control characteristic impedance and limit cross talk. The number of traces involved and the component density determines the number of board layers that are needed. If board size is not an issue then most boards can be designed using four layers.

The cost of the board is related to the number of layers that must be used. One way to add a few traces without adding layers is to break one of the conducting planes into two or more islands. Traces can then be run in the open area between islands. It is preferred to design boards so that traces maintain the same characteristic impedance over their entire length. If traces running between two conducting planes cross a gap on one of the conduction planes, the characteristic impedance at the gap will change. For many designs this may not cause a problem. For very short rise times, the resulting reflection at the gap can be an issue.

N.B.

When a trace crosses a gap on an outer ground or power plane, the return current must follow the edges of the gap. This is a classic radiating antenna structure. This is a bad practice for all logic. A path for the return current can be provided by adding a ground jumper next to the crossing trace. This is an improvement but the characteristic impedance will still not be well controlled in this area.

Ground and power planes provide return paths for transmission line current flow. Their primary function is not electrostatic shielding. On areas of a board where there are no traces, the ground plane or power planes is not needed for current return. It is important to keep these areas filled (flooded) with copper, as this provides mechanical and thermal stability for the board. This stability is needed during manufacture and board operation.

N.B.

Unused conducting surfaces should always be connected to ground or power. They should never be left floating. A floating conductor invites radiation and cross talk.

4.14   THE ANALOG/DIGITAL INTERFACE PROBLEM

Analog and digital circuits must often be located on the same board. Analog circuits can be sensitive to interference signals as small as 10 μV. Analog circuits can often rectify out-of-band signals and this introduces offset errors. An example of interference occurs in very accurate A/D converters. If the signal sample includes noise, the A/D converter generates an incorrect result.

It is often suggested that an analog circuit should have a separate ground plane. This solution still requires a connection between the two grounds. This solution forces any interfering currents to concentrate near the point of connection. This interference field, in turn, couples interference into the A/D converter. This approach is thus not recommended.

The easiest way to limit interference coupling is to make sure that the fields of the analog and digital signal processing do not share the same physical space. The rules that provide this separation are as follows:

Analog components should not mix with digital components.

Analog traces should not mix with digital traces.

Connections through a connector should be separated by function. This includes grounds and power. The circuits should not share the same power decoupling or power leads.

The only thing the analog and digital circuits can share is the ground plane. If this is done then there will not be an interference problem. The A/D converter should be oriented so that the analog input terminals face the analog circuitry.

N.B.

The key to interference control is field control.

4.15   POWER DISSIPATION

At high clock rates, it can be assumed that most of the logic traces will be series terminated in their characteristic impedance. Consider the series terminated case where current flows until the wave associated with each logic transition makes one round trip. The energy stored in the electric field is

$$E = \frac{nV^2}{Z} \times \frac{2d\varepsilon^{1/2}}{c} \tag{4.7}$$

where n is the number of logic transitions per second, V is the power supply voltage, d is the average trace length, ε is the dielectric constant, c is the velocity of light, and Z is the characteristic impedance of the average line. This energy is eventually dissipated in driver resistances, in termination resistors, in the dielectric, in the copper along the paths of transmission and finally in radiation. The energy lost to radiation is usually very small compared to the heating losses. For $n = 10^{12}/s$, $V = 3.5$ V, $d = 3$ cm, a board will dissipate about 10 W on logic traces. This is in addition to the dissipation in active components.

N.B.

Stored field energy is dissipated in terminations. When there are no terminations, stored energy is dissipated in radiation and IR losses from multiple reflections.

The heat generated in internal layers of the board must have a thermal path to outer conductive layers. If this is not done, the temperature rise on internal layers can cause the board to warp. If the outer layer is a ground or power plane then vias that interconnect planes can serve to provide a heat path. Heat sinks mounted directly on components may be necessary.

4.16   TRACES THROUGH CONDUCTING PLANES

When a trace crosses through a single conducting plane, the field associated with any signal must cross to the other side of the plane. The only path for the field is the hole (via) used by the trace. This transition will have very little effect on the transmission, as the characteristic impedance of the path will be fairly constant. When a trace crosses more than one conducting layer, the problem is more complex. The problem can best be viewed by locating the path the return current must take. This path will usually be through a nearby via that interconnects the two planes. The result is the field pattern that spreads out between the conducting planes, which means that there is a poorly defined stub associated with the transition of signal between layers. This spreading of fields can result in wave reflections and cross talk. To limit the problem, the vias used for the transition should be close together so that the space used by the fields is limited. Ideally, the path should be coaxial, but this is not practical with simple vias.

A trace that crosses more than one layer might be routed near power on one layer and near ground on another layer. The return path for signal current must cross in the power and ground plane space and through the nearest decoupling capacitor. The transmission line character of this capacitor is in series with the return path. This capacitor supplies field energy for many signals and can be a source of cross coupling. The field pattern in the ground/power plane space spreads out. If this space is shared by several traces in transition, there can be cross talk. An example of a via current path is shown in Figure 4.5.

The field is located between the forward and return current arrows just as it would be in any simple transmission line. There are other paths possible, but this is the path that stores the least field energy. This is the path that nature will use.

If the trace transitions between two grounds or two power planes then the field path will not involve a decoupling capacitor. Loops will still form that involves two vias, one for the signal and a second for the ground/power plane interconnection. The fields will still follow a path through holes in the planes. These separated vias will also disrupt the transmission line path used by the logic.

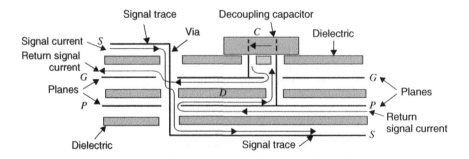

Figure 4.5 Current path for a signal when a via crosses layers. Note: the characteristic impedance is not controlled in areas C and D. Cross talk can occur in these areas. Area D can have a low characteristic impedance. Wave action takes place between the two current paths.

N.B.

For logic in the gigahertz range, vias that transition logic between different layers should be avoided.

N.B.

The return path for current supports the E and H fields needed for the forward transmission of a wave. There are two conductors and current flowing in both directions, but there is only one wave carrying energy in one direction.

4.17   TRACE GEOMETRIES THAT REDUCE TERMINATION RESISTOR COUNTS

Consider a wide trace that has a characteristic impedance of 25 ohms. If the line splits into two 50-ohm paths of equal length and the lines are unterminated, the reflections will arrive back at the split at the same time and the line will behave as one 25-ohm line. This configuration takes one series termination resistor.

This same principle can be used to form three or more logic paths. The length of each transmission path must be held constant. At each split, the parallel characteristic impedances must match. An example of a three-way split is shown in Figure 4.6.

There is a net saving in surface area as there are no spaces between traces over a portion of the run and there are fewer terminating resistors.

4.18   THE CONTROL OF CONNECTING SPACES

Traces over and between ground planes define the spaces used by fields to propagate signal or energy. These spaces must be controlled along an entire path or there will be reflections resulting in delays. A good design controls the spaces between conductors and not just the geometry of the traces. This control must be for both signal propagation and the supply of energy from decoupling capacitors.

A problem can occur when the path of energy flow must transition between traces and components or between layers on a circuit board. The problem of connecting to a decoupling capacitor has already been discussed. The problem of connecting to the pins of a component can also be an issue. If the field must cross from the outer surface of a pin to the inner surface to reach the die then the field must spread out. This can be viewed as adding inductance or more properly as a change in characteristic impedance, which results in delays caused by reflections. A transition that is only 1/16th inch long can cause trouble when the rise times are very short.

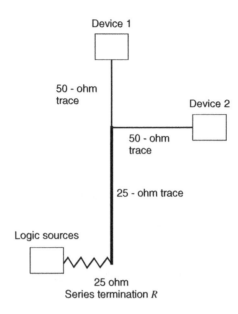

Figure 4.6   A branching transmission line. Note: The 50-ohm lines should be the same length.

A problem that is often ignored is the connections to a metal-cased power transistor. The only entry point for field energy is through the spaces around the pins. There is no way that field energy can transition through the metal enclosure. This includes energy to run the device and energy that leaves the device as signal.

The problem of supplying energy to a circuit from a low impedance source requires that the characteristic impedance be controlled along the entire energy path. To have low characteristic impedance, the spacing between conductors will usually be under 1 mil. This is true whether the path involves conducting planes, capacitors, DTLs, traces or coaxial cable. Controlling small dimensions at an interface poses a problem. This is a system problem as components must be compatible at the interface for both signal and power flow.

Interconnections where small spaces must be controlled requires compromises. Some misalignment must be permitted to make any connection at an interface practical. If the transition is limited to a few mils in length, the effect will not be noticed. Note that a length of 60 mils would be considered excessive. The example shown in Figure 4.7 shows a connection between a coaxial line and a component.

The edges of all connections must be chamfered to allow for misalignment. Note the way that the conductor geometry controls the space that is used by the fields.

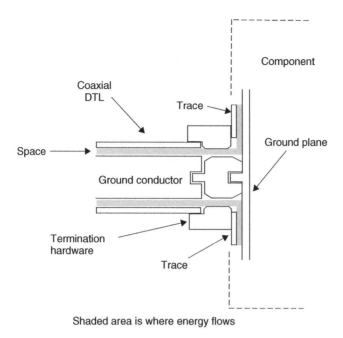

Shaded area is where energy flows

Figure 4.7 A coaxial connection to a component.

4.19 ANOTHER WAY TO LOOK AT ENERGY FLOW IN TRANSMISSION LINES

Consider a transmission line where the dielectric thickness is 0.01 cm. If the static voltage across the line is 1.0 V then the E field is 10,000 V/m. The energy in the space between the conductors is proportional to the product of the square of the E field and the volume of that space (Section 1.12). Now, consider a part of this line where the conductor separation is increased to 0.1 cm. In this section the E field is only 1000 V/cm. The energy density is reduced by a factor of 100, but the volume is increased by a factor of 10. This means that the static energy stored per unit length in this section is reduced by a factor of 10.

Consider a 0.1-V wave in the two sections of line. When the thickness is 0.1 cm the energy flow in a wave is lowered by a factor of 10. It would take 10 round trips for a wave of equal amplitude to move an equal amount of energy in the section with a thicker dielectric.

GLOSSARY

A/D Converter (Section 4.14): An integrated circuit that converts analog signal levels to digital signals.

Analog/Digital interface (Section 4.14): On circuits boards that contain analog and digital circuits, the interface occurs at the IC that handles both signal types. If the fields for the analog and digital signals use different volumes of space there will be no cross talk. This includes traces, components, connectors, and ground planes. Separate ground planes are never needed.

Coaxial (Section 4.5): A round outer cylindrical conductor geometry with a round center conductor. The outer conductor can be braid or a thin hollow metal tube. A coaxial cable is usually built to have a specific characteristic impedance. A low characteristic impedance requires that the conductor spacing be very small compared to the cable diameter.

Circuit board resonance (Section 4.6):
  1. *Circuit Sense.* The natural frequencies of the parallel capacitors used to decouple a circuit board. The series inductances of capacitors form resonant circuits that are all in parallel when the capacitors are paralleled. The resulting two-terminal impedance has a series of low and high values.
  2. *Digital Sense.* The back and forth movement of energy on an unterminated transmission line.

Decoupling transmission line (Section 4.4): A four-terminal capacitor where the electrical length of the transmission path is defined and the characteristic impedance at the load terminals is controlled in value.

Dielectric losses (Section 4.12): The heating of the dielectric that results when fields in the dielectric change amplitude.

Four-terminal device (Section 4.4): A capacitor with four terminals uses two to introduce energy and two to draw energy out the conductor geometry. Resistors can be built four terminals where two terminals connect to the source of current and two inner terminals are used to measure the voltage drop across the exact resistance specified.

Ideal voltage source (Section 4.2): A source of voltage that does not sag when any load is applied or removed. A zero impedance source.

Interposer board (Section 4.5): An intermediate connecting surface between a die and its mounting hardware.

Loop area (Section 4.2): The area between connecting leads that carry current. The area between a trace and a ground plane is a loop area. The area formed by the power supply leads to a die is a *loop area*. The connecting leads of a capacitor form a loop area.

Loss tangent (Section 4.12): When a sine wave field propagates in a dielectric, there is a phase shift that results from losses that is not associated with wave position. The tangent of this loss phase shift angle in 1 in of transmission is used as a measure of dielectric loss. The loss in decibel per inch is given by Equation 4.6.

Maxwell's equations (Section 4.11): A set of differential equations that describe all electrical activity in terms of fields.

Natural frequency (Section 4.7): The frequency where the reactance of the capacitance equals the reactance of a series or parallel inductance. The reciprocal of $2\pi(LC)^{1/2}$.

Port (Section 4.3): A terminal that carries a voltage or a signal into or out of a component or device. In theory, the port also includes the return conductor.

Power time constant (Section 4.2): The length of time it takes to increase the energy flow in a section of transmission line to within 37% of a final value.

Resonant circuit (Section 4.6): A parallel or series capacitor and inductor. Any active circuit that responds like a series or parallel inductor and capacitor.

Service entrance (Section 4.1): The utility power entry point in a facility.

Skin depth (Section 4.11): The depth where a sinusoidal current that flows in a conductor is reduced to 37% of its surface value. Skin effect has a complex effect on the character of the leading edge of a step function. In general, the rise time increases over distance.

Split ground (Section 4.13): When a conducting plane is not continuous under a trace, the plane is said to be split. The split is often used to allow one or more traces to reach an inner point or the split can involve islands of ground or power. These splits cause return current to move along the edges of the split. This pattern of current flow is classic to slot antennas. The slot causes wave reflections and radiation.

Wave (Section 4.1): The electromagnetic energy that flows along a transmission line after a switch closure connects a voltage to the line. Energy is reflected when a wave reaches a transition in characteristic impedance.

5

SIGNAL INTEGRITY ENGINEERING

5.1 INTRODUCTION

Logic circuits have become so complex that they cannot be memorized by a human being. The thousands of details that make up a board design are best handled by computers. An engineer is needed to decide on a design approach and to direct the activity of the computers. If the new board design is like an existing design, the designer can repeat an earlier approach. If a new set of specifications requires using faster or different integrated circuits, then the engineer must consider how the new performance requirements might change the board layout. This usually means starting over.

There was a time when components could be interconnected in any convenient way and the circuit would function. As the rise times have shortened, the problem of handling logic signals, clocks, and power distribution can no longer be left to chance. Today, every aspect of a design requires careful scrutiny. One logic line that is incorrectly connected to a stub can spoil an entire design. One overshoot can limit the reliability of a component. One circuit that is not correctly decoupled can cause a logic glitch. The problems that must be considered extend into board manufacturing.

Digital Circuit Boards: Mach 1 GHz, First Edition. Ralph Morrison.
© 2012 John Wiley & Sons, Inc. Published 2012 by John Wiley & Sons, Inc.

If the circuit boards warp or radiate, the design must be rejected. If the traces are too narrow or closely spaced, the board may be too expensive or not manufacturable.

It is not practical to let a computer simulate all performance aspects of complex circuit and include such diverse topics as cross coupling, overshoot, dielectric losses, reflections, radiation, temperature rise, and skin effect, as well as logic. Many of these factors can be grouped and handled separately by the designer. Computers can check for timing, logic, and wiring errors. The designer must make many initial choices and adjust the parameters until a design emerges. An important part of getting started is to generate a set of design rules that properly distributes the error budget for each type of signal. This process is the essence of signal integrity engineering. Design rules should be in place before any attempt is made to route a circuit board. When problems are encountered the design rules may have to be adjusted. An iterative process is usually needed before a design takes shape. Meeting performance requirements and meeting cost limitations while staying with a fixed board size may take several iterations.

In a design, the choice of IC family can make a difference. Smaller packages have fewer connections, which reduces the parasitic associated with internal lead lengths. Smaller packages also allow more trace room on the board, which reduces cross talk. If possible, select devices with longer rise times, as these devices are far less demanding in terms of engineering time.

5.2 THE ENVELOPE OF PERMITTED LOGIC LEVELS

Every logic type has a set of specifications that defines the voltage range for a logic 1 and a logic 0. This permitted range of operation includes temperature variations and parameter differences in manufacturing. If a logic level is outside of this specified range, the manufacturer does not guarantee performance. If a logic 1 is too low or a logic 0 is too high, the logic transition at clock time is undefined. If the logic 1 is too high or the logic 0 is too low, there can be a logic error or damage to the product. As an example, the envelope of permitted values for TTL logic is shown in Figure 5.1. TTL is used as an example even though this logic class is not in common use. Newer logic types will have different rise and fall times, as well as different operating voltages. The TTL logic shows the 0.7-V clamping levels when silicon diodes are used. Not all logic provides clamping diodes. In some high speed logic, Shottkey diodes are used where the clamping voltage is 0.3 V. Regardless of the logic class, the arriving voltage (logic) must lie between the two outer curves or the circuit may not function properly.

In this example, the nominal logic levels are 0 and 5 V. The upper limit for TTL is 5 V plus the breakdown voltage of the internal diode clamp or 5.7 V. The lower limit is 0 V minus the clamp voltage or -0.7 V. The inner voltage boundary is set at $+0.8$ and $+4.2$ V.

The envelope in Figure 5.1 includes the rise and fall times of the logic. The objective is to design the board, so the logic signals arrive at the component inside of this envelope at clock time.

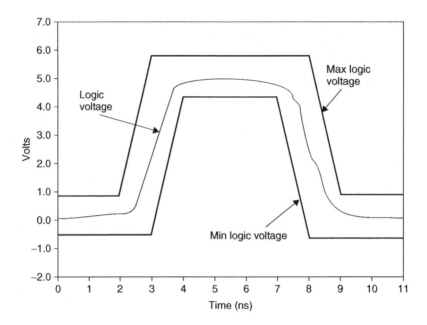

Figure 5.1　The envelope of permitted logic levels for TTL logic.

5.3　NET LISTS

In a typical circuit board design, many of the signal trace runs are very similar. As an example, consider traces where the logic drivers are the same, the trace length is between certain limits, the rise time is fixed, there are no stubs, the lines are series terminated, and the traces are all stripline with defined trace parameters. Traces that have this fixed set of attributes are said to belong to one net. In a typical design, there can be dozens of distinct nets. Computer programs are available that can verify that traces assigned to a net meet the requirements of that net.

The designer must first decide on IC types, board material, trace widths, etc. There are tools that will analyze transmission line performance for a set of choices. If a signal does not arrive within the expected time envelope then the trace may have to be assigned to a different or new net list. If the design reaches a snag and the design must be changed then all net lists must be re-examined.

5.4　NOISE BUDGETS

ICs with the same type number can vary in performance. These variations can result from differences in manufacture or from differences in application. These latter differences result from reflections, cross coupling, various voltage drops, or power supply fluctuations. The designer must make sure that these variations do not cause the logic signal to arrive outside of the control envelope. The limits of signal variation are called the *noise margin*. For the TTL logic shown above, the peak-to-peak noise margin is limited to 1.5 V. An allocation of error might be reflection ±0.05 V, cross coupling ±0.1 V, power supply sag 0.3 V, and various logic voltage drops 0.15 V. These numbers would then be used in the design of the circuit board. If the reflection error is too high then a line termination adjustment might be required. The reflection error in this example would include permitted variations in the characteristic impedance of traces.

In analyzing a noise budget, losses subtract from a nominal signal. Cross coupling can add or subtract from the logic level, so it is necessary to assume that it is a worst case number. For example, cross coupling can come from both sides of a trace.

A list of the logic errors is given below. The important errors are discussed in the following sections. Each error must be interpreted as a signal voltage. They are as follows:

1. Driver variations, voltage, and source impedance
2. Characteristic impedance variation and effect on arriving voltage
3. Effect of nearby traces on characteristic impedance
4. Terminating resistor accuracy
5. Cross talk, forward and backward modes, one or both sides of a trace
6. Voltage drops, line and ground, skin effect. "Ground bounce"
7. Common-mode coupling
8. Rise-time effects
9. Stubs, both mid line and at the ends of the line
10. Power supply variations (peak-to-peak ripple)
11. Dielectric losses.

5.5 LOGIC LEVEL VARIATION

A part of the noise budget is taken up by the inability of a logic driver to provide a voltage that reaches the limits of the power supply. The limits of voltage swing are called out in the manufacturer's specifications. For a logic 1, the highest loaded voltage is specified as VOH_{MAX}. The lowest maximum loaded voltage is VOH_{MIN}. The highest logic 0 is VOL_{MAX}, which defines the other limit to the logic voltage swing. The difference between VOH_{MIN} and VOL_{MAX} defines an envelope that must be considered in handling a noise budget. When parallel terminations are used the terminating resistor can be made intentionally high to compensate for a lack of voltage swing. In series terminations the resistor can be lowered to accommodate for the lack of voltage swing.

Series line terminations must take into account that the driver may have a finite source impedance. The source impedance plus the series resistor must match the line impedance. If the sum of the resistances is high then the logic level at the open end of the line will be low. If the sum of the resistances is low then the logic level at the open end of the line will be high. These adjustments are sometimes needed to keep errors within the assigned budget. In fast circuits, the location of the termination resistors is critical. Ideally, series resistors should be placed at the driver. Any lead length ahead of the resistor is a short transmission line that causes reflections that in effect adds to the rise time. If there is doubt about the permitted trace length then some sort of simulation is suggested. As stated earlier, a resistor is really a short section of lossy transmission line. The characteristic impedance of the line is modified by changes to the trace geometry required to mount the resistor.

If there are routing difficulties, the design rules may have to be adjusted. Making traces narrower may increase cross talk, reduce reliability, and make it difficult to meet the characteristic impedance requirements. Meeting all performance requirements and staying within cost constraints is often a difficult engineering problem.

5.6 LOGIC AND VOLTAGE DROPS

The signal levels that arrive at a logic gate can vary for a number of reasons. We have already considered the voltage drop in the driver logic. In Section 4.12, we discussed how skin effect increases the trace resistance. The voltage drop in traces can best be understood by first considering the resistance of a bar of metal.

The dc resistance of a bar of metal having a cross-section A and a thickness h is $\rho s /A$, where s is the length of the bar and ρ is the resistivity. If the bar of metal is a square then the square has a resistance of

$$R = \frac{\rho}{h} \qquad (5.1)$$

When the thickness h of the bar is small compared to width and length, the conductor takes on the form of a sheet of metal. When the sheet of metal is used on a circuit board, it is referred to as a *ground* or *power plane*.

A sheet of metal can be characterized by the resistance of a square of that material. The term used is ohms-per-square or Ω^2. The assumption that must be made is that the current flows uniformly across the square between opposite edges.

N.B.

A square of material has a resistance between opposite edges that depends on thickness and not on the size of the square. If two squares are placed in series and the current flows in the series connection, the resistance simply doubles. If the current flows across the width the resistance is one half.

The term *two-ounce copper* means that two ounces of copper are deposited on 1 ft^2 of material. Two-ounce copper has a resistance of 24.2 $\mu\Omega^2$ at dc. At 10 kHz, the resistance doubles to about 48.4 µΩ. Above 10 kHz, the resistance increases proportional to the square root of frequency. At 1 GHz, the resistance increases by a factor of 316 to about 0.015 Ω^2. For a trace that is 5 mil wide and that is 1 in long, the resistance is computed by considering that there are 200 squares in series. The resistance is therefore 3.0 ohm. Because skin effect keeps the current near the surface, the resistance at 1 GHz for 2 oz or 1/2 oz copper is essentially the same.

The current on the ground plane spreads out depending on the height of the trace above the ground plane. For a 5-mil trace, 10 mil above the conducting surface, the width of the current path in the ground plane is approximately 15 mil. The resistance of this return path will be about 1.2 ohm/in. Thus, the total resistance in the signal path would be about 4.2 ohm/in. A trace that is 10 mil wide will reduce this resistance to about 2.5 ohm/in. Since skin effect is proportional to the square root of frequency, the resistance at 100 MHz would be reduced by a factor of about 3.

N.B.

For 3-V logic, the current in a series terminated 50-ohm line is 30 mA. The voltage drop in 2.5 ohm is 75 mV.

N.B.

Consider a trace routed over a conducting plane. At high frequencies, the return current flows under the trace and not in the entire plane. Skin effect limits the depth of penetration in this conducting plane.

The resistivity of copper is 1.72 µΩ-cm. This is low compared to solder plating that has a value of 13.5 µΩ-cm. For plated traces, skin effect forces current to use the plating, which raises the resistance. To reduce this resistance, it is possible to plate outer layer traces with gold. Note that gold has a resistivity about equal to copper. Gold plating has been used in analog (microwave) circuits, but it is not generally used in digital designs. If the traces are embedded in a dielectric then plating is not needed (embedded microstrip or stripline). The areas that must still be plated are through-holes, pads, and test points.

5.7 MEASURING THE PERFORMANCE OF A NET

When a trace configuration is open to question it may be necessary to construct the net and measure its actual performance. As the rise times get smaller, the difficulty in making useful measurements increases. The rise time of the measuring hardware becomes an issue and also the techniques used by the operator. An oscilloscope probe is really a field measuring device. Where the probe ground is connected makes a big difference. The loop areas formed by the probe at the point of measure must be kept very small. It is necessary to verify that no signal is sensed when the probe is connected to the probe ground at the point where it connects to the board. If there is signal then the loop formed at the tip may be excessive. It is also possible that current in the probe shield may be the source of error. If there is direct probe coupling then a better probe may be necessary.

It is a good idea to test a net without its terminating load (terminating gate). Clamping action obscures overshoot on any arriving signal. When the gate is removed, a small capacitor may have to be used as a dummy termination. In high speed circuits, the loop area formed by this termination is important. If the connection is too long, it amounts to a stub on the transmission line. Any terminating capacitor should be a surface mounted component to keep the loop area as small as possible.

N.B.

If actual measurements of a net are not practical, then it may be necessary to use a field solver to determine performance.

5.8   THE DECOUPLING CAPACITOR

In Chapter 3, it was pointed out that a decoupling capacitor and its connection can be considered as short sections of transmission lines. The connections that are made inside and outside of the IC are also transmission lines. There are usually several parallel connections to logic traces in an IC driver, as shown in Figure 5.2.

When the logic switch closes, the process of obtaining energy involves transmissions and reflections along all the connected transmission lines. It is important to note that the initial energy will come from the electric field associated with the die and connected logic and power traces. The next nearest source of energy is the ground/power plane capacitance and finally energy will come from the decoupling capacitor. If the energy in this geometry is adequate and if there is sufficient time to gather energy through multiple reflections, a logic signal of sufficient amplitude will propagate over the connected transmission lines and arrive at the logic gates within the time permitted by the error envelope. If the energy used in this transmission is not immediately replaced then the transmission at the next clock time will have a lower voltage.

The path that fields must take to supply energy to the decoupling capacitor is worth discussion. The energy stored in the ground/power plane field cannot cross through a conducting plane. The field carrying this energy must cross through holes and gaps in the ground plane. In Figure 5.2, the field must fold around the gap between the embedded trace and the ground plane. The field on the top of the board

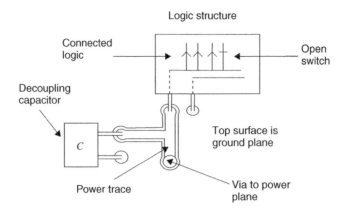

Figure 5.2 A set of connections at a logic structure and a decoupling capacitor.

is poorly contained. This could be considered an inductance or a section of high impedance transmission line. When a wave finally reaches the capacitor, the reflected wave carrying energy must follow the same path in reverse order. Many round trips are needed to supply the needed energy. In a circuit sense, the field above the ground plane represents inductance. In a transmission line sense, the transmission path has a segment of high characteristic impedance.

N.B.

This single demand for energy may involve over 100 reflections and transmissions. It does not take a computer analysis to show that this is a poor layout.

N.B.

Fields flow in spaces and not in traces. The geometry of this space needs to be controlled if performance is to be improved.

Decoupling capacitors have natural frequencies that are usually well below 1 GHz. A 1000-pF capacitor with a series inductance of 0.1 nH has a natural frequency of 159 MHz. A step function with a 100 ps rise time has a rise-time frequency of $1/\pi\tau_r$ or 3.18 GHz. These frequencies are over an order of magnitude apart. In a circuit sense, the capacitor has a series inductance that slows the flow of energy.

The ground/power plane has a power time constant of about 1 ns. During this first nanosecond, the voltage will sag at the point of demand. If a DTL (decoupling transmission line) is used, it will be the first element to supply energy after a logic request. The only delay that remains is the connection from the logic switch to the DTL. If a DTL is a part of the IC package, then even this delay can be eliminated.

N.B.

Ripple on a power supply for one component will be noise for a logic line that arrives from another component.

5.9 CROSS COUPLING PROBLEMS

In high speed logic, a major consideration is cross coupling. This topic was covered in Chapter 3. For logic signals with short rise and fall times, the major contributor to cross talk is the backward coupled wave. This wave lasts for the full round trip of the culprit wave and can cause the victim's logic level to be outside of its operating envelope. To do a full analysis of cross talk, a field solver may be needed.

Every logic transition requires the supply of field energy. As rise and fall times get shorter and clock speeds rise, the need to reduce power consumption follows. Reducing operating voltages reduces power consumption and makes it possible to raise the clock speeds. Lowering the operating voltage means smaller logic swings. This, in turn, reduces the available noise margin for the logic. Unfortunately, the need for noise margin increases with shorter rise and fall times. At shorter rise times, there is increased cross talk and greater signal loss caused by skin effect. Another element in the error budget that may have to be considered is dielectric loss. This error can be reduced by using a higher grade of board material.

The simplest way to reduce cross talk is to increase the spacing between traces. The simplest way to limit skin effect losses is to widen the traces. Both of these changes are in the direction to increase the size of the board. To avoid making a larger board, the only approach left is to increase the number of trace layers. Adding trace layers is in itself not sufficient. It is necessary to have a conducting plane associated with each added trace layer or it is impossible to limit crosstalk and control the characteristic impedance of the traces. Obviously, these added layers increase the cost of the product. Fortunately, there is another factor that should be considered in the cost equation. When more functions are handled by the ICs, the effect is to reduce trace count and board size.

5.10 CHARACTERISTIC IMPEDANCE AND THE ERROR BUDGET

A part of the error budget must include errors caused by variations in characteristic impedance. When series terminations are used, the signal level and the termination resistor must match the line impedance. A mismatch produces an error signal. There are many other factors that influence the characteristic impedance of a trace. The obvious factors are variations in trace width, thickness, and spacing to the nearest conducting plane(s). The secondary effects that should be considered include the presence of nearby parallel traces. Nearby traces include side-by-side traces or traces on a nearby layer. Another source of error is the variation in dielectric constant across the board. Circuit board manufacturers are able to control the characteristic impedance of traces to within ±10%. This may only be the impedance of a test strip and not the characteristic impedance of all traces. Test strips can also be applied to inner layers. It is up to the designer to request these strips. It is also up to the designer to interpret the manufacturer's claims correctly.

Terminating resistors may have a tolerance of about ±10%. The driver source resistance and its tolerance must also be considered. When the errors related to resistor tolerance and characteristic impedance are equal, the logic level at the load can vary a maximum of ±20%. Resistors that have a 20% tolerance will obviously use up more of the error budget. The effect of parasitic capacitance across the resistor was discussed earlier. To limit any resulting overshoot, a longer resistor is preferred over a stubby resistor. Unfortunately, a longer resistor alters the characteristic impedance of the transmission line path. In high speed circuits, the choice of resistor type should not be left to chance. Tests should be made to verify performance. The resistor type and the mounting treatment must be specified so that performance is repeatable.

Designers must be careful to specify all the materials used in the fabrication of their boards. There are different weaves of glass that can be impregnated with epoxy. This can have an effect on the way the dielectric constant varies across the board. If the error budget is tight then variations in board material could cause logic errors to exceed the budget. All of the board material detail used in a final product should be a part of the specifications. This way the board can be manufactured by a second supplier and the performance can be duplicated.

It is standard practice for board manufacturers to measure the characteristic impedance of test traces at 1 MHz. If the clock rates in the circuit are at 1 GHz, the dielectric constant at this frequency will be lower. As an example, the dielectric constant can drop 15–20% between 1 MHz and 1 GHz. Note that the characteristic impedance varies inversely to the square root of the dielectric constant. If the dielectric constant drops 16% in this range then the characteristic impedance rises 4%. This is the value that must be used in any error analysis.

Narrow traces are often specified to limit the layer count. The narrower the trace the more difficult it is to control characteristic impedance. This is related to the dimensional tolerances that can be maintained by the board manufacturer. As pointed out earlier, closer traces can also increase cross talk. Before narrower traces are accepted, a very careful net analysis is required.

Dielectric losses must be considered especially on longer trace runs. These losses tend to increase the rise time. The designer may want to test actual hardware to see how significant the signal delay, signal loss, and crosstalk will be. These tests can place limits on trace length to keep the signals within the error budget.

5.11   RESISTOR NETWORKS

Multiple terminating resistors are often supplied as a single in-line package or SIP. When resistors are used for series terminations, separate connections must be made to each resistor. Note that on a SIP package, a separate lead goes from the top of the resistor to the board for each resistor. To reduce lead length in short rise-time circuits, a DIP or dual-in-line package is preferred. Any short transmission line between the signal source and the resistor can add to the signal propagation time. For a SIP package, the ends of the resistors nearest the board should connect directly to the logic. This is another case where a field solver may be required to determine if the performance is adequate. In general, high speed circuits are easier to manage if the resistors are mounted individually and not in SIP or DIP packages. There are solutions where the resistors are buried in the board as a part of vias. Another solution involves surface-mounted components. The approach used is a cost versus performance issue.

When parallel terminations are used, a group of resistors are often connected together at one end of the package. This common tie is then connected to the logic reference conductor, which is often the ground plane. The traces carrying the signals should terminate on their loads (logic gates) before reaching the resistor. If the termination is made in the reverse order, the result will be a short open-ended stub that connects to the load. Under these conditions, the voltage will attenuate at the resistor and double at the load and continue to reflect. The voltage at the load will be uncertain. This problem is shown in Figure 5.3.

A problem exists when a group of terminating resistors shares a common tie. Assume that the common connection to the resistors is brought out at one end of the SIP. When all but one of the logic lines are at logic 1, the current in the common lead can be quite high. This problem is shown in Figure 5.4.

The voltage drop across this common impedance is an error signal. This error must be considered in any error budget analysis. To reduce the coupling, the ground

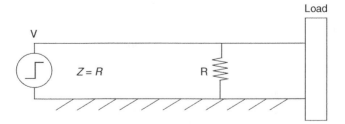

Figure 5.3   The location of parallel terminations.

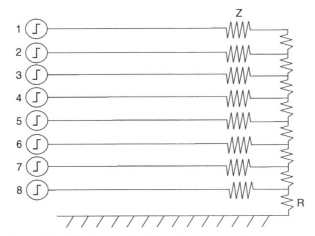

As each logic line is energized the voltage across the
common resistors R increases. Each Z is 50 ohms. Each R
is in the milliohms. The R to common is apt to be the largest.

Figure 5.4   Common impedance coupling in resistor packaging.

connection can be made in the center of the SIP. The best solution is to connect all
resistors to a conducting bar that bonds along its entire length to the ground plane.
This technique eliminates any common impedance coupling.

N.B.

The connections to terminating resistors are transmission lines. If not handled
correctly, the signal error budget can be exceeded.

5.12   FERRITE BEADS

Engineers frequently attempt to add ferrite beads to power traces to reduce ripple.
As we have stressed, ripple results because the demand for current requires many
reflections. The bead approach is similar to that used in the low frequency filtering of
analog signals, where RC or LC filters are used. Ferrite filters fail in most cases. First,
the inductance is very small and second, raising the transmission line impedance at a
point can only add another point of reflection.

A single-turn ferrite inductor might have an inductance of a few tenths of a
nanohenry. At 1 GHz, the reactance is around 1 ohm. This is much lower than the
characteristic impedance of most transmission lines. If a shunt capacitor is added, it
is obviously in the right direction. The capacitor alone might be of some help. If

multiple turns are used on the bead, the resulting parasitic capacitance will lower the natural frequency of the inductor to well below 1 GHz.

Ferrite materials have very little permeability above a few megahertz. The impedance they offer is in the form of losses. In fact, the inductance is usually not much different from the same air-core geometry. In high current applications, any core material will probably saturate. Air-core inductors do not saturate.

The presence of a ferrite bead changes the geometry of a transmission line usually in the direction to increase the characteristic impedance. This is in the direction to add reflections and increase the noise on the line.

If the problem is cross talk then the beads may do nothing more than space the conductors. In this case, the core material can be air that is, of course, much less expensive.

N.B.

Ferrite beads do not limit interference on digital transmission lines.

5.13   GROUNDING IN FACILITIES: A BRIEF REVIEW

N.B.

A grounding rod eliminates interference the way a flag pole eliminates sunlight.

The term *grounding* has many meanings. In utility power, the term means a connection to earth or its equivalent. In an aircraft or an automobile, the metal framework is the equivalent of earth. In a building, one of the power conductors at the service entrance must be earthed to provide a path for lightning. If lightning should hit the utility wiring, the intent is to provide an earth path for current flow outside of the facility. For uniformity, these grounding rules are followed even if the power is brought into a facility underground.

The grounding electrode system in a building includes such diverse items such as building steel, plumbing, gas lines, conduit, metal floor mats, boilers, motor frames, air ducts, and all equipment grounding conductors. In effect, all of the conductors in a building that could contact a power conductor must provide a fault path back to the service entrance ground. There can only be one grounding electrode system in a facility. The grounding electrode system can be multiply earthed. The grounding electrode system cannot be used as a path for power current.

Equipment grounding conductors (safety conductors, bare wires, or green wires) are used for fault protection and for shock avoidance inside a facility. Equipment grounding conductors are a part of the grounding electrode system. Equipment grounding conductors must be installed to carry any fault currents back to the grounding point at the service entrance. To keep the inductance in the fault path low, the equipment grounding conductors must parallel the path taken by the power conductors. Equipment grounding conductors must be run inside the same conduit as the power conductors. All metal enclosures that could contact a power conductor must be connected to an equipment grounding conductor. A fault to this conductor must trip a breaker. In very special cases, a temporary fault condition may simply cause an alarm rather than disrupting the power. Examples might be in electric furnaces or on assembly lines. This approach is permitted if an electrician is on duty at all times.

N.B.

The grounded power conductor or neutral must be earthed once in a facility. This point is at the service entrance. The grounded power conductor or neutral cannot connect to any equipment ground at any other point in a facility. This rule must be followed, if the fault protection system is to function properly.

A facility can have several sources of separately derived power. In these systems, there can be separate neutrals, feeders, and equipment grounding conductors. Examples are auxiliary generators, special distribution transformers, or battery backup systems. Separately derived power sources are treated like a new service entrance. Each separately derived neutral is connected once to the nearest point on the grounded electrode system of the facility. Equipment grounds associated with this power are returned to this same point.

A connection to earth is rarely below 10 ohm. If there were two such connections, the total resistance might be 20 ohm. Assume that a facility has two earth grounds (isolated from each other). If a hot lead somehow connects to earth, the earth resistance is placed across the power line. This would allow 6 A to flow. In most branch circuits, this current is not high enough to trip a breaker. This means that somewhere in a facility, there could be "grounded" conductors that are physically close together but that are 120 V apart. A person touching these grounds could be electrocuted.

N.B.

Because of the safety danger, the code prohibits the use of two grounding electrode systems in one facility.

The impedance between all grounded (power) conductors must be in the milliohms at power frequencies. It is preferable to bring separate sources of power into a facility in the same physical area.

5.14   GROUNDING AS APPLIED TO ELECTRONIC HARDWARE

In electronic hardware, a conducting plane is often called a *ground*. This is not a power industry term. In some analog designs (below 100 kHz), there may be several different grounds or reference conductors. These grounds or commons may or may not be directly earthed. In general, these conductors are used in the processing of non-power-related signals. When these reference conductors are inside of hardware, they are not controlled by the National Electrical Code (NEC). In commercial products, this grounding is controlled by UL or its equivalent. The terms *signal common* or *signal reference conductor* are preferred over the term *ground*.

A cell phone may have a conducting plane used as a return path for forward transmission. Obviously, this conductor is not grounded or earthed in any way. This plane is certainly not controlled by the NEC. These "grounds" control the flow of field energy within the device. In some ways, it is unfortunate that we use the same word to describe all of these different conducting surfaces. The semantics problem gets even more complicated when we add adjectives such as signal ground, digital ground, power ground and output ground, and safety ground. These adjectives have little effect on making a facility electrically quiet.

N.B.

Nature does not read the labels we place on conductors. She also does not read color codes.

The ground conductor used in a digital circuit is used as a current path (return path) for logic signals. Correctly stated, the common conductor is used to confine and route the fields used to carry logic information and energy to the various logic components. The fields are confined in the space between traces and conducting planes. In a digital circuit, the ground plane is one of the signal conductors. External to hardware, grounds should not be used to carry return signal current. In cases where analog signals are carried over long distances, shielded cables should be used. The signal return may be connected to the shield at the source end of the signal. It is important that interference currents flow in the shield and not in the grounded signal conductor located within the shield enclosure.

In analog designs below 100 kHz, several signal reference conductors are often used. In low frequency analog designs, there is no economic advantage in using a ground plane. Circuits can be made to work with or without this added plane. The reasons relate to the limited bandwidth, and the fact that there are very few analog signals processed at the same time in one circuit.

Analog circuits can sometimes be impacted by the presence of fields that are above the band of interest. In these cases, there may be signal rectification in the circuitry that results in in-band signals. Once the rectification takes place there is no way to eliminate the error. This means that passive filtering may be needed to keep out-of-band signals from reaching a sensitive input circuit.

In analog board layouts, the signal fields and the interfering fields can usually be kept apart without the need for conducting planes or even shielded conductors. In digital work, ground planes are the best way to tightly control the fields associated with logic and energy transport. Fortunately, logic signals have amplitudes greater than 1 V and a 0.1 V error is not usually an issue. In the analog world, it is often necessary to worry about microvolts of interference. This means that the two approaches to interference control are very different even though they both involve fields.

Equipment grounds of any sort should not be used as a signal return conductor. The reason is obvious. Conducting surfaces outside of the hardware must be power grounded (earthed) to satisfy the requirements of the NEC. These conductors are associated with power safety and therefore with power line filters. Since these conductors are exposed to the outside world, fields in the local environment reflect and couple to transmission lines formed by these conductors. These fields include switching transients as well as television, police radio, radar, and cell phone signals to mention a few sources. Reflections of waves at these conducting surfaces imply surface currents. Mixing these interference signals with signals of interest is not a good idea. The approach often used is to place all signal conductors that interconnect hardware inside of shielded cables. The shields carry the interference currents on their outer surface. These shields are often bonded (grounded) to the housings at both terminations. This shielding arrangement can be deficient in noisy environments. In a quiet environment, unshielded ribbon cable can be used. An example might be a connection when the cable is routed inside of a rack enclosure.

In a typical cable connection, some of the signal conductors inside the shield are connected to the circuit commons in both pieces of hardware. This means that there is a parallel path for interference currents. Some of the interference follows the shield and some of it follows the common conductors inside of the shield. If the cable shield is of adequate quality and if the shield terminations at the hardware are properly handled, most of the interference current will stay on the outside of the shield.

Another way to limit the flow of interference currents is to use optical couplers or coupling transformers. The latter solution requires balanced drivers, balanced lines, and balanced receivers. If optical couplers are inside of the receiving hardware then the entering cables can still couple interference into the hardware. Optical coupling must be placed at a shield boundary, if it is to be effective in limiting interference coupling.

N.B.

To avoid interference, shield currents should flow onto outer conducting surfaces.

It is poor practice to float any circuitry inside of hardware in an attempt to control interference coupling. The best approach is to use balanced or differential circuitry, so that any interference coupling is symmetrical. This way, balanced circuitry can be used to reject the interference as a common-mode signal. Fortunately, digital signals have high noise immunity. In very sensitive low frequency analog work, guard-shield techniques can be used. In this solution, the guard shield grounds at the signal source and does not connect to the amplifying hardware at low frequencies. The shield protects the signal along its entire route, including the path inside the receiving hardware. The input connections must be near the connector. Above 100 kHz, the guard shield is best connected to ground at both ends of its run. This can be handled by placing a series RC connection between the shield and signal common. A typical connection might be a series 100-ohm resistor and a 0.01-μF capacitor.

N.B.

Interference currents should not flow in signal conductors. For this reason, shields should not be used to carry signal currents. There are a few exceptions such as short runs of microphone cable.

Conducting planes play the same role inside or outside of hardware. Inside of hardware, they are used to confine signal fields. Here, the plane can be the return path for signal current. Outside of hardware, the conducting plane cannot be used as the return path for signals, as this same conductor carries interference currents. The only role the external conducting surface can have is to limit the coupling of external fields to the cable shields routed on its surface.

Coupling is proportional to the loop area between the cables and the conducting plane. In most applications, openings in the shield coverage (lack of bonding) at the connector create the biggest problem. If the fields associated with shield currents penetrate the hardware at the connector then the presence of ground planes is of little value.

N.B.

Ground planes do not stop interference. They can only reflect fields with a horizontal E field component.

Some of the current on the outside surface of a braided shield can cross over the inside surface of the braid. This means that there are fields coupled to the inside of the cable. The ratio of external current to coupled signal voltage is called a *transfer impedance*. Fine braid is better than coarse braid. Twin braid is somewhat better and solid-wall shielding is the best. Manufacturers of cable can supply figures on transfer impedance for various cables.

5.15   INTERNAL GROUNDING OF A DIGITAL CIRCUIT BOARD

A question that is often asked is, where should a digital ground plane be connected (grounded) to the hardware housing? The first observation is that a metal enclosure plays a small role in controlling the fields associated with the logic. This control exists entirely with the logic ground plane and the routing of traces. It is still a good idea to connect the enclosure to the circuit ground plane, as allowing it to float raises all sorts of new issues. This connection should be made to limit equipment ground currents from flowing in the ground plane of the circuit board. This can be done by connecting the housing to the circuit common, where the secondary of the power transformer connects to the circuit board ground plane. This practice is shown in Figure 5.5.

N.B.

Grounding a circuit board does not attenuate the fields that might exist near the circuit board. The ground connection might change the field configuration in areas near the connection. Multiple grounding of a circuit board ground plane to the housing is not recommended, as it invites interference currents to flow in common conductors on the board.

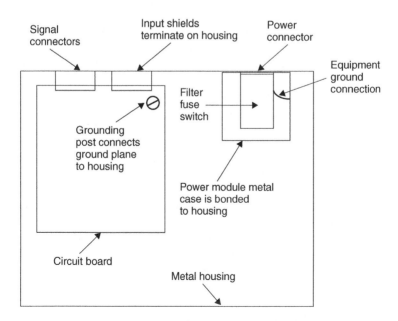

Figure 5.5   Grounding a housing to a circuit board.

N.B.

When practical, board connectors should be located in one area of a board. This keeps interference currents that flow in cable shields from flowing across the board.

5.16   POWER LINE INTERFERENCE

Power line filters can keep some of the common-mode and differential interference from entering the hardware. In most cases, available filters do not function well below a few megahertz. The filtering reference conductor is usually the equipment grounding conductor that carries filter current back to the service entrance. If the equipment grounding path is inductive (long) then the filter cannot be effective. It is important to note that even with a good filter installation, some of the equipment ground current uses the racks, the signal conductors, and the cables that interconnect hardware to return to the service entrance. These currents represent field energy in the facility.

In a single phase power service, the neutral or grounded power conductor carries power currents for all the loads on a feeder and its branch circuits. This means that this conductor is not at the same potential as the equipment grounding conductor, which does not carry load current. The voltage impressed across a power transformer primary has some common-mode content caused by this neutral voltage drop. The content depends on the attenuation provided by the line filter. Filtering effectivity depends on whether every power conductor is filtered and how the filters are installed.

If there is a power transformer, there is a capacitance from turns on the primary coils to the secondary coils. Common-mode interference on the primary coil will flow through this capacitance to the circuits on the secondaries. The paths involve all conductors that connect to circuit common. This means that there is interference current flowing on signal commons that exit the hardware and any equipment ground connections. These signal commons are often shields or conductors in a ribbon cable.

It is against the code to place any component in series with the equipment grounding conductor. This includes filter inductors. If this component should open circuit then the fault path is broken, creating a hazardous condition. The equipment grounding conductor must connect to the conducting hardware framework. If the connection is made to an inside surface then fields carried on the conductor can directly radiate into the hardware. The right way to terminate the equipment ground on hardware is on an external conducting surface.

It is difficult to build switching circuits without generating some radiated field. This field can couple to power leads and exit the hardware even if there are line filters. In hardware that uses DC/DC converters, the fields that couple out on the power line can be a serious problem. These converters involve semiconductor switches that have short rise times. The currents that flow in a few picofarads of parasitic capacitance can be hundreds of milliamperes. This current can be reduced by careful construction inside the converter transformers and by the proper placement of converter filters. In large systems, there can be hundreds of these converters in operation, all switching at different frequencies. The result can be a very noisy facility.

Large motors often use speed or torque controllers that switch power in mid cycle. This switching can create interference on the power conductors. This can be troublesome, if this same power enters other electronic hardware. The fields resulting from this switching can radiate and leave the conductors. These fields cause surface currents on all nearby conductors and can enter hardware on shields and signal conductors. Fluorescent lamps create fields that can propagate on all nearby conductors. These currents can also enter hardware on signal cables and on power conductors.

5.17   ELECTROSTATIC DISCHARGE

Electrostatic discharge (ESD) is a source of interference that must be addressed in every design. ESD is a potent pulse of electromagnetic energy that is characterized as being 5 A with a rise time of 1 ns. This pulse of current creates a field that can destroy nearby components. The rise-time frequency is about 300 MHz. The near-field/far-field interface distance is about 16 cm. Near the pulse, the wave impedance is low and the wave energy is dominated by the magnetic field. Circuit damage is most likely when a field couples to a loop area that involves an input gate.

In most logic structures, the loop areas are very small, so it is difficult to couple any significant voltage into a surface-mounted IC. The larger loop areas are apt to involve decoupling capacitors, connectors, or other component connections. Note that a coupled voltage of 0.3 V is all that is needed to cause a clamping diode to conduct, and this could result in a logic error or even component damage. If the current pulse were to flow in a ground trace then common impedance coupling could add voltages that would destroy a component.

The first line of defense is to keep the pulse from using conductors in the circuit board. If the pulse can hit a cable, then the cable must be shielded. If the pulse can follow a circuit conductor then some sort of diode protection is needed to divert the current so that it does not enter a signal path. As an example, keyboards should have a conducting membrane to divert current to a shield structure.

N.B.

It is a good idea to view this problem in terms of field patterns. Conductors provide a path for fields to follow. Conductor geometry should be used to keep interfering fields from using the space around signal conductors.

In floating circuits, the effect of an ESD hit is to add charge to the entire structure. This charge will usually bleed off in air. If a transformer is used to charge a battery then some sort of high impedance path should be provided, so that any build up of charge can bleed off to earth around the transformer. This path can be 100 MΩ. If there is a local shield and there is no bleed path, then the circuit should be built so that any eventual arcing will take place in the transformer and not in the circuit.

The ESD field can enter into a metal enclosure through an aperture. A half wavelength at 300 MHz is about 50 cm. A 1-cm aperture will attenuate the field by a factor of 50, and this is usually not good enough. For this reason, apertures of all types should be closed by using gasket material or wave guide construction. Fans can use honeycombs to attenuate fields and still allow air passage. See "wave guide" in the "Glossary" of this chapter.

ESD pulses that flow along a shielded cable can couple into the signal leads. This coupling is related to the transfer impedance of the cable. The most likely coupling point is at the cable connector. In critical applications, connectors with back shells should be used.

To calculate the voltage induced into a loop near the pulse, first calculate the H field intensity by noting the distance from the current. Convert this H field to B field by using the permeability of free space as a multiplier. Then calculate the flux by multiplying by the loop area. The induced voltage is the time rate of change of this flux. As an example, consider a distance from the ESD pulse of 0.1 m. The H field is $5/0.1 \times 2\pi$ A/m. The B field is $4\pi \times 10^{-7}$ H or 10^{-5} T. If the loop area is 0.0001 m^2, the flux is 10^{-9} lines. If this flux changes in 1 ns then the voltage in the loop is 1.0 V.

Zappers are pulse generators that are used in testing hardware. The pulse amplitude can be varied from about 1 to 15 kV. There are usually two modes of operation, a direct contact and a near contact. In the direct contact mode, a pulse of current is injected into a structure without an arc. In the arcing mode an external field is generated that can enter through apertures. The most critical arcing, voltage is about 7 kV. Above that voltage, most of the energy is lost in heat and light. Testing should proceed starting at a low voltage and progress through all key points in both modes. When trouble is encountered, the testing is stopped and needed changes to the structure are made.

GLOSSARY

Apertures (Section 5.16): Any opening that allows electromagnetic energy to enter a conductive enclosure. There is no attenuation of an external field (worst case) when the aperture's maximum dimension is one-half the wave length. The field attenuation for a smaller opening is the ratio of half wave length to aperture opening.

Apertures that are spaced so that surface currents can circulate freely around the aperture are independent. The field penetration for several independent apertures is additive. Dependent apertures such as screens, seams, or ventilation arrays act as one aperture.

Coax: See shielded cable.

Common (Section 5.11): A term used to mean ground or the conductor at zero volts. This conductor is sometimes called the *reference conductor*. Terms such as signal common, output common, digital common, and power supply common are often used.

Cross coupling/cross talk (Section 5.4): The coupling of wave energy from one trace to a nearby trace. The coupling involves both forward and reverse traveling waves. Coupling only occurs where the wave amplitude is in transition. Reverse wave coupling causes most interference problems. The amplitude depends on rise time. The reverse wave lasts twice as long as the coupling time. The forward coupled wave is a pulse that increases in amplitude with time. Forward coupling is apt to be a problem on long outer traces.

Embedded microstrip (Section 5.6): The outer layer traces are covered by a dielectric. These traces are not plated with solder so the skin effect resistance is that of copper and not solder. Forward wave cross coupling is reduced, because the E field is reduced in the dielectric.

Envelope (Section 5.2): A graph of permitted values. The graph usually shows the maximum and minimum permitted values as a function of time or of position.

Equipment ground (Section 5.12): An NEC term. Any conductor that could come in contact with a "hot" or ungrounded power conductor. All equipment grounds must be bonded together to provide a short and immediate low inductance return path for a fault current so that a breaker will interrupt the power. Equipment grounds can be multiply connected to earth.

Electrostatic discharge (Section 5.17): The arcing that results when an insulated object accumulates a charge large enough to ionize the air. This charge results in an electric field that surrounds the object. When this field builds up on a human and the human gets near a grounded metal object a pulse results that discharges the accumulated charge.

ESD: See electrostatic discharge.

Ground (Section 5.12): In power terminology, this word means an earth connection or its equivalent. In circuit terminology, it often means the low or common side of a power supply.

Grounding (Section 5.12): The connection between a circuit common and a larger external conducting surface such as a metal housing, a rack, or the earth.

Grounding rod (Section 5.12): A conductor buried in the earth to make an electrical connection to the earth. Typical connections to earth are about 10 ohm at low frequencies. In the power industry, this conductor is called a *grounding conductor*. It is used to provide a lightning path for the neutral or grounded conductor at the service entrance to a facility.

Grounding rods do not control electromagnetic interference. They are required by code to provide lightning protection and electrical safety.

Guard shield (Section 5.13): A shielding method used in analog instrumentation to protect very small analog signals carried over long cables. This shield is connected to signal ground at the signal source. It is brought into the instrumentation to shield the signal, but it is not connected to the input circuitry. The common-mode signal that is rejected is derived from the signal pair. The guard shield is terminated on output ground external to the instrumentation at frequencies above 100 kHz.

NEC (Section 5.13): The National Electrical Code. This code is law in most cities in the United States. It represents permitted practice in power wiring for facilities. The code protects against electrical shock, fire, and lightning. There are many ways to build a facility and have it satisfy the code. All interference problems can be resolved without violating the code. The words used in the code are defined in the code. Power engineers use these definitions.

Net list (Section 5.3): A list of traces on a circuit board that have similar character-istics.

Noise budget (Section 5.4): The range of values a signal can take and still allow a logic circuit to perform correctly. The budget includes items such as cross talk, errors caused by variation trace characteristic impedance, variations in terminating resistors, line driver limitations, and temperature rise.

Ohms per square (Section 5.6): The ratio of current to voltage drop in a square of thin conducting material when the current flow is uniform between opposite edges on one surface. The resistance is independent of the size of the square.

Resistor network (Section 5.11): A group of resistors packaged for simple mounting on a circuit board. The packaging can be SIP or DIP. In high speed logic, the lead lengths and spacing of these structures can be a problem.

Separately derived (Section 5.12): A power term used to describe a power source that has its own neutral. Examples are a distribution transformer or an auxiliary power source. A computer power center or CPC has its own distribution trans-former with a grounded neutral. In all cases, the added neutral is connected once to the one grounding electrode system of the facility. One advantage to using a separately derived system is that it allows a second neutral conductor not used by other hardware in a facility. Neutral voltage drops can be a source of common-mode interference that enters electronics through the power transformers.

Shielded cable (Section 5.14): Shielded cables are conductors routed inside of a conducting sheath. Cables that are used to transport signals at frequencies below 100 kHz are often covered by a conductive braid. To control characteristic impedance the outer conductor must maintain shape and the conductor spacing and dielectric constant must be controlled. Single conductor cable intended for high frequency work is called *coax*.

Shielding effectivity: The ratio of performance before and after the shield is included in the design.

Test strip (Section 5.10): A trace added to a panel of circuit boards used to test characteristic impedance.

Transfer impedance (Section 5.13): The transfer of external field energy into a cable. It is measured as the ratio of external surface current to voltage measured at the output terminations for a unit length of cable. Since transferred energy travels in both directions, the voltage used in this ratio is double the signal measured at one termination.

The shielding effectivity of braid can be negligible at frequencies above 100 MHz. Two shields or very dense braid can be an improvement. The lowest transfer impedance is found in thin wall solid conductor shielding. This tubing can be corrugated to improve flexibility. The wall thickness can be just a few mils to be effective electrically.

Wave guide (Section 5.17): A hollow cylindrical conductor that can support wave transmission without a center conductor. The lowest frequency that can be transported has a half wavelength equal to the aperture opening. For frequencies lower than this value, the wave guide attenuates electromagnetic field energy exponentially. The attenuation in decibels is $30d/h$, where d is the depth and h is the dimension of the maximum opening. This attenuation is in additional to the aperture attenuation. See aperture.

Wave guide construction (Section 5.17): Apertures or seams that have depth. An example might be a folded conducting lip or a honeycomb.

Zapper (Section 5.17): A test device that generates ESD pulses.

6

CIRCUIT BOARDS

6.1 INTRODUCTION

Early circuit boards were made of bakelite, where components were soldered on to mounting pins and vacuum tubes were plugged into sockets. Insulated wires were used to interconnect sockets, pins, and components. Building a board was labor intensive. Component leads were cut and bent to provide strain relief, interconnecting leads had to be stripped and tinned, and all the parts had to be assembled and soldered by hand.

A great step forward was made when copper/epoxy laminates were introduced. Traces replaced insulated leads. Components were mounted in holes that were plated and tinned. If shielding was needed, then critical circuits were surrounded by metal boxes. The logic was slow enough that circuit traces were routed point-to-point, and there were few cross talk or overshoot problems. Wire-wrap technology was introduced when the number of interconnections could not be handled by traces on two layers. Progress in IC design then increased clock rates, which meant that rise and fall times were shorter. Attempts to operate at higher clock rates were often met with failure, as there were just too many problems in signal integrity. It was soon recognized that wire-wrap technology was a blind alley, and it was necessary to go to multilayer boards with added ground planes to provide circuit performance. The trend to smaller

Digital Circuit Boards: Mach 1 GHz, First Edition. Ralph Morrison.
© 2012 John Wiley & Sons, Inc. Published 2012 by John Wiley & Sons, Inc.

trace widths and multilevel boards soon followed. We often take for granted the amazing technology that we now enjoy. It makes our computers, servers, and cell phones possible.

Two-layer circuit boards are in common use even today. These boards are copper-clad laminates composed of a glass epoxy core with copper plate on both surfaces. The expression "two-layer board" means that there are two conducting surfaces. The board manufacturer drills holes for vias and mounting components and etches the copper to form traces and component pads. The manufacturer then plates the holes with copper and solder plates holes and pads so that components can be mounted and soldered to the board. The traces on this board must connect the power supply and common to every active component, as well as interconnect all of the signals.

Today, integrated circuits perform many tasks. This requires a large number of pinouts on individual ICs. A large pin count means that many interconnecting traces are required. For many commercial products, four-layer boards are the preferred solution (four conducting surfaces). Even larger trace counts have made designers turn to smaller trace widths and closer spacings. In this approach, conducting planes and traces are intermixed on each layer. Any unused areas on a layer are "flooded" with copper connected to ground or power. Flooding is a good idea, as it provides a more robust board, reduces board warpage due to heating, and helps in the radiation of heat.

The number of transistors on one die can now exceed 1.8 million. To achieve this component density, the individual transistors have been made smaller. This has resulted in devices that switch in a shorter time. The present 90 nm technology will soon be even smaller making the rise times even shorter.

The issues of cross talk and radiation are closely related to rise and fall times and not clock rates. This one fact has a significant impact on circuit performance. To illustrate the problem, assume that a driver transitions in 100 ps. If energy to drive the connected logic traces is not immediately available, the voltage to the device will sag. This can have two effects. The logic can malfunction, and there can be increased radiation from the board. The problem increases when a number of signal transitions occur at the same clock time.

There are techniques that can be applied even to a two-layer circuit board that allows it to handle fast logic. A practical solution to this problem is discussed in Section 6.7.

6.2   MORE ABOUT CHARACTERISTIC IMPEDANCE

Signal integrity in high speed digital design is closely related to the control of characteristic impedance. As rise and fall times get shorter, the problem of control becomes more and more critical. At clock rates of 1 GHz, all logic lines must be either series or parallel terminated to avoid overshoots. Stubs of any length must be avoided, as the delays that result can impact performance.

There are many variables that affect the characteristic impedance of traces, which cannot be easily put into an equation. For example, the published equations relate to rectangular traces. In practice, traces have rounded edges and copper etching and plating modifies trace width and shape. The characteristic impedance is also affected by the presence of nearby traces. Resistors that are used for line terminations have parasitics that influence their performance. The presence of a resistor modifies the trace spacing, and this affects the characteristic impedance. Terminations are often capacitive, and this modifies the nature of any reflection. In outer layers, the need to deposit copper in drilled holes causes added copper to be deposited on traces. This added thickness modifies the characteristic impedance. Unused portions of a via that extend through a circuit board can act as a stub. Removing this unused copper adds board cost.

Dielectric constants vary from point-to-point across a board. One cause for this variation is that the dielectric constant for the glass weave is different from that for the epoxy. For loosely woven glass, it is possible that some traces may lie on top of the weave and others may lie on top of epoxy, and this varies between boards. In designs where the characteristic impedance must be closely controlled, the laminate should use closely woven thin threads of glass. The weave that is chosen must be a part of the board specification. Even with a fine weave, there can be variations across the board.

There is a spectrum associated with step functions and the dielectric constant; and thus, the characteristic impedance of traces varies with frequency. The board manufacturer may check the characteristic impedance of a test trace to verify board uniformity but remember that this is probably a test of a single trace and this trace is probably on an outer layer. Furthermore, the test may be made at 1 MHz and not at 1 GHz. Of course, the real test occurs when the board is manufactured and put into operation.

In general, published equations for characteristic impedance have been simplified so that they solve a typical or standard problem. The user should find out the range the parameters can take before using the equation. In spite of the many problems, characteristic impedance equations contain a lot of information and provide a starting point in any design. The document IPC 2251[1] is frequently referenced in the literature. This document gives approximate formulas for the characteristic impedances of microstrip and stripline traces in FR-4. These are the equations we will use. This document can be purchased from www.ansi.org.[2]

There are many web sites that will calculate the characteristic impedance of various trace geometries. Unfortunately, the equations in use are often hidden and their limitations are not always stated. They can be used as a starting point but additional checking is always advised. A comparison between published equations will show that they may not be in agreement. An equation often has a "sweet spot," where it is

---

[1] IPC stands for Institute of Printed Circuits. It is a global organization headquartered in Lincolnwood, IL, USA. It maintains specifications and procedures, as they relate to circuit board manufacture. The organization also holds classes on various aspects of board manufacture.

[2] ANSI stands for American National Standards Association. See www.ansi.org.

most accurate. Because of these uncertainties, the most reliable approach is to use a field solver that makes use of Maxwell's equations.

6.3 MICROSTRIP

Microstrip refers to the traces on the outer layers of a circuit board. Some drawings show a trace resting on top of the dielectric of a given thickness. Others show the trace as partially embedded in the dielectric. The characteristic impedance Z_0 for the microstrip, as shown in Figure 6.1, is given in Equation 6.1:

$$Z_0 = \frac{87}{\sqrt{\varepsilon_R + 1.41}} \ln \left(\frac{5.98h}{0.8w + t} \right) \tag{6.1}$$

This equation will give good results for $0.1 < w/h < 3$ and where $\varepsilon_R < 15$.

It is interesting to hold the characteristic impedance constant and note the relationship between trace width and dielectric height. Figures 6.2–6.4 show curves of constant characteristic impedance, where the dielectric constant is 4. If the dielectric constant is 3.5, the characteristic impedance rises 5%.

Note that the equations are continuous, but the user faces many dimensional restrictions. For example, trace thickness standards are half 1- and 2-oz plated copper. This thickness will increase on outer layers when copper is plated in drilled holes. In practice 1/2-oz copper ends up thinner than the expected 0.7 mil. The useable thickness is between 0.5 and 0.6 mil. This reduction in thickness results from cleaning operations during manufacture. The designer should consult the manufacturer to determine the final trace thickness after manufacture.

Dielectric thicknesses are limited to the laminates or cores that are available. The equations provide some insight into just how thick the dielectric must be to bring the characteristic impedance into range. Here again, there are many thicknesses available, and there are also many tolerance problems. It is wise to work with a manufacturer and use materials that he understands and that he keeps in stock.

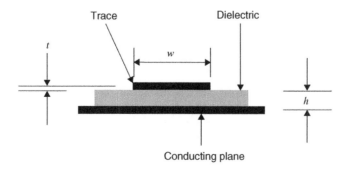

Figure 6.1 Microstrip geometry.

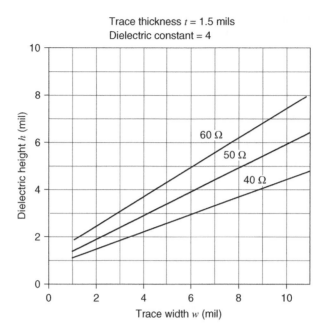

Figure 6.2 Microstrip. Trace width versus the dielectric height needed to hold the Z_0 constant. Trace thickness equals 1.5 mil.

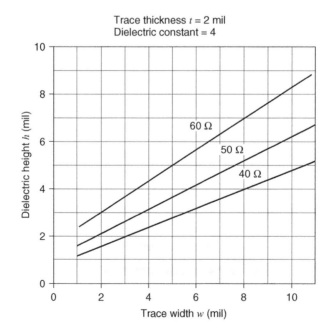

Figure 6.3 Microstrip. Trace width versus the dielectric height needed to hold the Z_0 constant. Trace thickness equals 2.0 mil.

Trace thickness $t = 2.7$ mil
Dielectric constant $= 4$

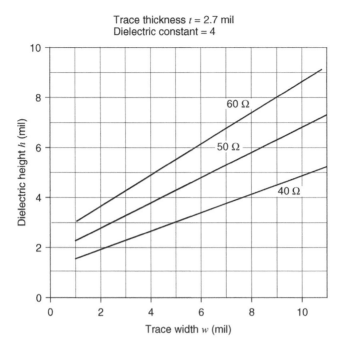

Figure 6.4 Microstrip. Trace width versus the dielectric height needed to hold the Z_0 constant. Trace thickness equals 2.7 mil.

The basic relationship between trace width, trace thickness, and dielectric height for 50-ohm lines and a dielectric constant of 4 is shown in Figure 6.5.

6.4 CENTERED STRIPLINE

Stripline refers to the traces that are located between conducting planes in the inner layers of a multilayer circuit board. The characteristic impedance of centered stripline for the trace geometry is shown in Figure 6.6 and given in Equation 6.2:

$$Z_0 = \frac{60}{\sqrt{\varepsilon_R}} \ln \left(\frac{1.9(2h + t)}{0.8w + t} \right) \tag{6.2}$$

where $0.1 < w/h < 2$, $t/h < 0.25$, and $\varepsilon_R < 15$.

It is interesting to plot the relationship between trace width and dielectric thickness. These curves are shown in Figures 6.7–6.9. Notice that the trace thicknesses are limited to the copper clad and copper foil thicknesses that are available and to the thickness after cleaning or plating.

Note that the dielectric spacing for a 50-ohm line are about double than those of microstrip. Notice that the characteristic impedance does depend on trace thickness. This dependence is shown in Figure 6.10.

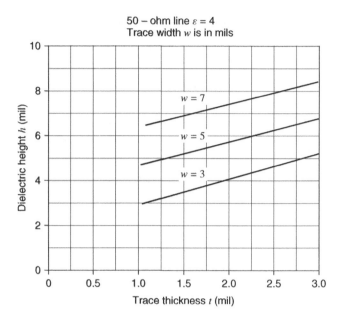

Figure 6.5 Curves showing the relationship between trace thickness t, trace width w, and dielectric height h for 50-ohm microstrip lines with a dielectric constant of 4.

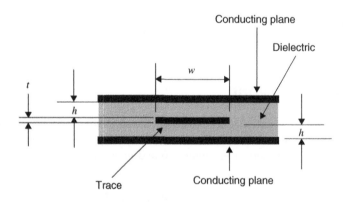

Figure 6.6 Centered Stripline.

6.5 EMBEDDED MICROSTRIP

There are several benefits to be gained when outer trace layers are embedded in a dielectric. Traces are not plated when the holes are plated. This allows tighter control of characteristic impedance. The reactive terms contributing to cross talk are more apt to cancel. This reduces the amplitude of the forward cross-coupled wave. The traces

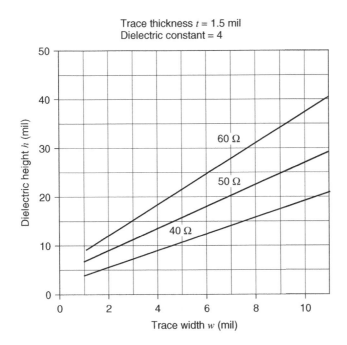

Figure 6.7 Centered stripline. Curves of fixed characteristic impedance versus trace width and dielectric height for a trace thickness of 1.5 mil.

are not plated, so surface currents flow in copper and not in solder. This reduces transmission line losses due to skin effect.

The geometry of embedded microstrip is shown in Figure 6.11.

The equation for the characteristic impedance of embedded microstrip is

$$Z_0 = \frac{56}{\sqrt{\varepsilon'_R}} \ln \left(\frac{5.98h}{0.8w + t} \right) \tag{6.3}$$

where $h_1 > 1.2h$ and $\varepsilon_R < 15$. The effective dielectric constant ε' is

$$\varepsilon'_R = \varepsilon_R \left(1 - e^{-1.55h/h_1} \right) \tag{6.4}$$

For 2 mil of dielectric over the trace, the effect is to lower the characteristic impedance by about 1%. Refer to Figures 6.2–6.4 for curves that show how the geometry must change to keep the characteristic impedance constant.

6.6 ASYMMETRIC STRIPLINE

In some layups, there are two trace layers between conducting planes. These trace layers are usually routed at right angles to each other to reduce capacitive cross coupling.

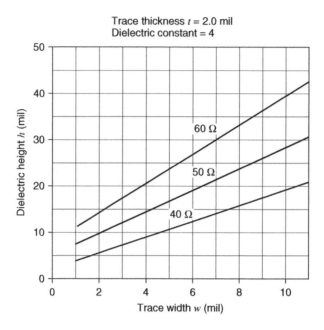

Figure 6.8 Centered stripline. Curves of fixed characteristic impedance versus trace width and dielectric height for a trace thickness of 2 mil.

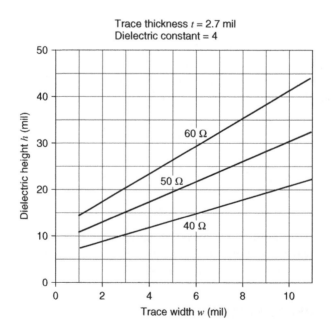

Figure 6.9 Centered stripline. Curves of fixed characteristic impedance versus trace width and dielectric height for a trace thickness of 2.7 mil (2-oz copper).

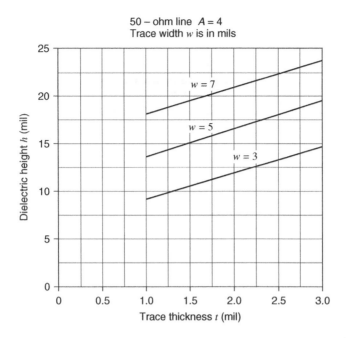

Figure 6.10 Centered stripline. Curves showing the relationship between trace thickness, trace width, and dielectric thickness for a 50-ohm line and a dielectric constant of 4.

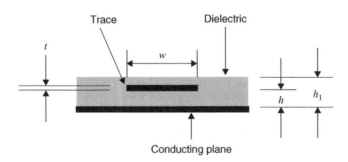

Figure 6.11 The geometry of embedded microstrip.

The characteristic impedance of traces in one layer is not significantly affected by the crossing of traces on the second layer.

In some layups, a single trace layer is placed between conducting planes with a different dielectric thickness under and over the trace. The formula for the characteristic impedance is given by Equation 6.5. The conductor spacings for this configuration

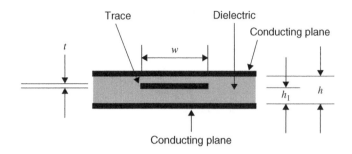

Figure 6.12 Asymmetric stripline.

is given in Figure 6.12.

$$Z_0 = \frac{80}{\sqrt{\varepsilon_R}} \ln\left(\frac{1.9(2h + t)}{0.8w + t}\right)\left(1 - \frac{h}{4h_1}\right) \tag{6.5}$$

where $h_1 > h$, $0.1 < w/h < 2$, and $t/h < 2.5$.

6.7 TWO-LAYER BOARDS

In conventional construction, a two-layer circuit board does not have a ground or power plane to form transmission lines. This means that a signal trace that interconnects components must have a nearby ground or power trace to form a controlled transmission line. For a logic and ground trace running in parallel on one surface, the characteristic impedance will usually be above 70 ohm. This results because the edge capacitance per unit length is a small value. There are several ways by which the capacitance can be increased to reduce the characteristic impedance, such as if the logic trace is placed between two ground/power traces, if the traces are thicker, or if the trace spacing is reduced.

The trace geometry we will consider for a two-layer board is a logic trace between two ground or power traces. The largest part of the field energy is concentrated in the space between the inner trace edges. We will consider trace widths and spacings of 10 mil.

A possible construction for a fast double-sided board is to run traces left to right on the top surface and top to bottom on the reverse surface. The ground and power traces serve to connect power to the components and to form transmission line for each signal. It is good practice to use vias to interconnect ground and power traces on the two surfaces at regular intervals. These connections in effect form a pseudo ground/power plane for the board. Decoupling capacitors should be placed near the power and ground pins of each active component. Figure 6.13 shows a typical trace pattern.

The traces carrying power act as grounded traces as far as the characteristic impedance is concerned. When a logic signal propagates on the three conductors, the

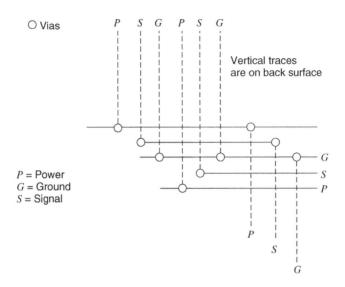

Figure 6.13 Trace pattern for use on a two-sided circuit board.

logic return current flows on both the power and ground traces. This means that the characteristic impedance is not a function of the static potentials on the traces.

The characteristic impedance Z_0 of a logic trace between two grounded traces is given approximately by Equation 6.6:

$$80\varepsilon^{-1/2} \ln \left(\frac{3s}{(0.8t + w)} \right) \tag{6.6}$$

where s is the spacing between traces, t is the trace thickness, and w is the trace width. The number of traces per linear inch is limited by the diameter of the via holes and the area of the surface pads that must be provided. Via hole diameters that are about 1/6th the board thickness are practical at little added cost. If the board thickness is 0.062 in then the hole diameter can be 10 mil. The pad diameter for a 10-mil hole is typically 25 mils, which means that trace spacing can be about 10 mil. If the board thickness is reduced then smaller diameter vias can be used.

For via pad diameters of 25 mil, trace spacing of 10 mil, trace widths of 10 mils, and a trace thickness is 2.7 mil, the characteristic impedance in air is approximately 73 ohm. If the traces are routed on a dielectric then half the field is in the dielectric. This means that about two-thirds of the transmitted energy is transported in the dielectric. The capacitance of the space in the dielectric is proportional to the relative dielectric constant. This means that if ε_R is 4 then the characteristic impedance of traces in this space is half that of air. The characteristic impedance of three traces over a dielectric can be considered to be $2Z$ and Z in parallel or $2/3Z$. In our example, this is an impedance of 48 ohm. If the trace thickness is 2 mil, the characteristic impedance is 50 ohm.

If the microstrip is coated with a thin dielectric, the traces need not be solder plated. This means that the surface currents will flow in copper that has a lower resistance than solder.

N.B.

Transmission lines can carry more than one signal in either direction. A trace pair can serve to direct the flow of operating power and the flow of a logic signal.

In board areas where there are no logic signals, ground and power traces can still be provided. An alternative is to flood these areas with copper and connect this copper to ground and power.

Vias are used to connect the logic traces in any required xy (Manhattan) pattern. The right angle bends in the transmission lines (signal path) will not introduce any significant reflections (Figure 6.13).

This connection geometry illustrates a basic problem in high speed logic. There is no way to reduce the loop area formed at the IC by the logic and ground connection. The pad and beam geometry forms a loop area that represents an uncontrolled section of transmission line. This loop area exists for all logic lines. Note that these loops are a source of radiation.[3]

Adding a ground pin next to every logic pin seems like a poor suggestion. The pin counts are quite high and rising. There still might be an answer if every pin is considered a two-conductor transmission line. Imagine a pin made of a stiff plastic strip with a plated conductor on each wide surface. One side would be logic and the other side would be ground. When the pin is installed in the board, the transmission line connection is automatically made.

If the traces carrying logic in the example mentioned above are separated by about 40 mil, there is the possibility of 25 traces per inch on each side of the board. For a 10 × 10 in board, that amounts to 500 traces running 10 in. If 30% of the board is taken by pads and 70% of the remaining board is useable, there can still be about 245 traces running 10 in. If the average trace run is 4 in, there can be 612 logic traces. If 70% of these trace runs are useable then the trace count drops to 428.

If a high trace count is not needed then the trace widths and spacings in the grid structure can be scaled up. Wider traces are in the direction to reduce both skin effect losses and cross coupling. A typical construction is shown in Figure 6.15. A typical signal termination is shown in Figure 6.14.

---

[3]The tradition of connecting conductors to form circuits is not an easy one to change. We are used to the idea of connecting traces as transmission lines on the board but at the IC package the concept often stops. The logic connections and the ground connections are often widely separated. In fast logic, this causes reflections (delays) and radiation. Ideally, the characteristic impedance of every transmission line should be controlled up to the die itself.

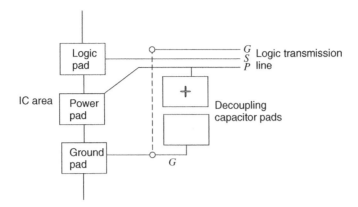

Figure 6.14 Treatment of the logic trace at a termination.

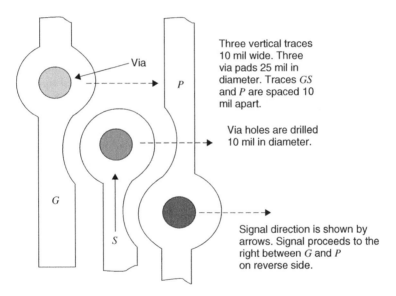

Three vertical traces
10 mil wide. Three
via pads 25 mil in
diameter. Traces GS
and P are spaced 10
mil apart.

Via holes are drilled
10 mil in diameter.

Signal direction is shown by
arrows. Signal proceeds to the
right between G and P
on reverse side.

Figure 6.15 Via patterns for a logic transition.

6.8 FOUR-LAYER CIRCUIT BOARD

Multilayer boards are usually built using symmetric layer stacks. The core or middle
two layers consists of a copper-clad laminate. The thickness and type of epoxy-glass
used in the core, as well as the copper clad thickness must be specified by the designer.
In a four-layer board, the outer two conducting layers are copper foil separated from
the core by sheets of material known as *prepreg*. Prepreg is a partially cured layer of
woven glass and epoxy. Under heat and pressure, the prepreg bonds the outer copper

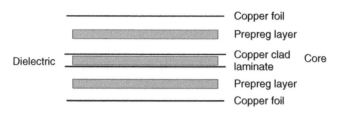

Figure 6.16 A four-layer board layup.

layers to the core. To get the required thickness and to eliminate the chance of shorts, several layers of prepreg are used. This stacking arrangement is called a *layup*. A four-layer layup is shown in Figure 6.16.

The outer layers of copper and prepreg add about 0.015 in to the thickness of the board. The remaining board thickness is in the core. Typically, the total board thickness will range from 0.040 to about 0.061 in.

There are several ways to use the four layers of a board to handle logic and power. Layers can be dedicated to logic traces, to ground, to power, to a mix of logic traces and power, or to a mix of logic traces and ground. The choice that is made depends on whether ground/power planes are used for decoupling and if the characteristic impedance needs to be controlled. If traces are limited to inner layers, there will be less cross talk, and these traces will not contribute to board radiation. The spacing required for trace routing between the two inner or core layers limits the amount of energy that can be stored for decoupling. If the outer layers are dedicated to ground or power then trace connections to components must be made using vias. Figure 6.17 shows two four-layer design configurations in common use.

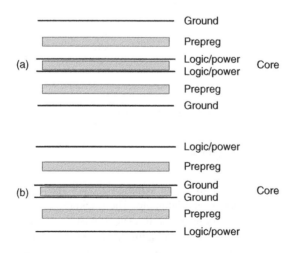

Figure 6.17 Two four-layer board configurations.

If the center conductors are both ground planes, the center laminate does not store field energy. For this reason, the choice of laminate material is not critical.

For board layers where power or ground planes are mixed with logic signals, care must be taken that all logic transmission lines are continuous. Breaks in the return path create reflections and affect operation. The rules developed in Chapter 2 must be followed or the board design will be compromised.

N.B.

As stated earlier, ground planes are not shields. They are used to define the paths taken by the flow of energy, where signals or power are transported by transmission lines. Do not break the return pathway of any signal.

A four-layer board typically accommodates two trace layers and two ground planes. Islands of ground or power can supply some of the required decoupling energy. Reducing trace width and spacing can be used to increase the trace count. This approach will add to the board cost if the manufacturer has a high rejection rate. Also, narrower traces and closer spacings can increase the level of cross coupling, which impacts signal integrity. If these solutions are not practical then more trace layers are required.

N.B.

Six-layer boards are required when the number of traces exceeds the available surface area and a larger board is not practical.

Areas that have few traces and pads will receive the most plating. This modifies the characteristic impedance depending on trace location. For this reason, traces should be placed on inner layers when the characteristic impedance can be more tightly controlled.

The thickness of copper plating of through holes is critical if press-fit components are to be used. Manufacturers will often add copper pads or copper surfaces to a board to even out the plating thicknesses. These added pads are called *thieving* copper, as they rob plating from nearby conductors. When the characteristic impedance needs to be tightly controlled, it is best to use inner layers and avoid this plating problem.

6.9 SIX-LAYER BOARDS

Six-layer boards can be constructed by adding outer layers of copper foil to a four-layer layup. The layers of prepreg that bond the foil layers control the dielectric

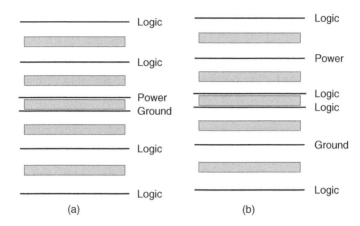

Figure 6.18   Layup designs for six-layer boards that should be avoided.

thickness. Another construction is to build the core from two clad laminates separated by prepreg and then add the outer layers of foil to form the layup.

A six-layer board allows for layups with adjacent ground and power planes. This geometry provides a distributed decoupling capacitor for the board. The closer the energy sources are to the loads the faster the energy can be supplied when there is a demand. The vias that connect this energy to the load must be closely spaced. This topic was covered in detail in Section 2.14.

Trace layers must be near conducting planes for effective control of the characteristic impedance. It is preferred to have a ground or a power plane on each side of a trace layer. Examples of six-layer layups that can be a problem are shown in Figure 6.18.

In these layups, the ground/power planes are spaced, so they do not supply much energy to the logic. In Figure 6.18a, the outer logic layers are not near a ground or power planes so that cross talk is likely.

In logic designs where the rise and fall times are short, the problem of cross talk is present. In general, cross talk is reduced for traces that are between the conducting planes. For this reason, it is desirable to place shorter trace runs on outer layers and longer trace runs on inner layers between the conducting planes. In high speed circuits, it is desirable to use embedded microstrip for outer layers. This reduces the level of forward wave cross coupling. This point was covered in Section 3.11. Two recommended layups for six-layer boards are shown in Figure 6.19.

"Flooding" areas with copper to form ground or power planes is always recommended. A ground or power plane will reduce radiation or cross talk only where the plane forms a transmission line return path for traces. In other areas, it makes the board more robust and helps to radiate heat.

The flooded areas that are under (over) traces will affect the characteristic impedance of nearby traces. Placing conducting planes near a part of a trace run can

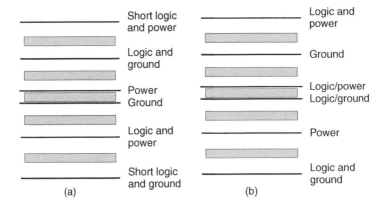

Figure 6.19 Acceptable layup designs for six-layer boards.

create problems if the characteristic impedance varies along the trace run. Obviously, power planes and ground planes that are not closely spaced may not contribute much energy for use in decoupling the board.

Traces that transition between layers must receive special attention. The return current must also transition so that the characteristic impedance of the transition is controlled. It is important to realize that fields cannot cross through a conducting plane except where there is a hole. If the transition causes significant reflections then the result is a problem in signal integrity.

GLOSSARY

Asymmetric stripline (Section 6.6): Traces between conducting planes that are not centered.

Centered stripline (Section 6.4): Traces that are on an inner layer of a circuit board that are evenly positioned between two conducting surfaces.

Clad laminate (Section 6.8): A glass epoxy layer with copper bonded to one or both surfaces.

Core (Section 6.1): The laminate structure used at the center of a circuit board. It is often glass epoxy with thin layers of copper bonded to the two surfaces.

Layer stack (Section 6.8): The set of laminates that make up a circuit board.

Layup (Section 6.8): The layers in their right order that make up a final circuit board.

Manhattan (Section 6.7): The xy pattern of traces similar to avenues and streets in the New York City.

Maxwell's equations (Section 6.2): A set of differential equations that describes all field phenomena in electricity.

Oz copper: A thickness of copper. When two pounds of copper are plated on 1 ft^2 of material, the copper thickness is 2.7 mil. This is called *2-oz copper*. The final thickness can change during manufacturing.

Prepreg (Section 6.8): Partially cured glass epoxy. When the layer stack is heated under pressure, the epoxy cures bonding laminate surfaces together.

Thieving (Section 6.8): In plating, areas with less copper are apt to receive an excess of plating. Adding pads of copper to a surface tends to even out the plating. The pads steal plating material from nearby points on the board.

ABBREVIATIONS
AND ACRONYMS

New acronyms appear in the literature each year. As time goes by, many of these acronyms fall into disuse. In a few cases, meanings may differ depending on the industry where they originate. It is possible for one abbreviation to have several meanings. When this happens, the reader must rely on the intent of the author. As an example, pH can mean hydrogen ion concentration or picohenry. The correct meaning should be obvious in context.

Authors often coin an acronym in an article when a word group is repeated many times. Sometimes, an acronym is used in one company, but it is not used by the industry. These acronyms will not appear in this list.

The following list includes the common abbreviations used in circuit board design, mathematics, electrical engineering, and physics. Fortunately, acronyms that survive rarely conflict with scientific or engineering abbreviations.

Abbreviations that do not conform to present day usage or good practice are not listed. Here are a few of the problem areas. The first letter in many engineering measurement terms is an "m". In an abbreviation, "m" could stand for milli, micro, or mega. It is good practice to use capital M for mega, small m for milli, and the Greek μ for micro. The letter m can also stand for meter, milliwatt, and mile. In the abbreviation dBm, the m means milliwatt. If m means meter, it can usually be inferred by context.

In an old copy of "The Radio Engineers Handbook" by Terman, the term mc is used as the abbreviation for megacycle. Today, the accepted abbreviation is MHz and it is read megahertz. We have come a long way in standardizing abbreviations.

Proper names used in engineering abbreviations require a capital letter. As an example, the A in mA stands for Andre Marie Ampere the French physicist. When the word ampere is used in a sentence, it is not capitalized. "ma" is not found in this list.

This list contains most of the common abbreviations used in this book and in circuit board design. It is not a complete list. It shows how extensively we use abbreviations and acronyms. It is hard to remember them all. If you do not find an abbreviation, the internet will usually provide some help.

Digital Circuit Boards: Mach 1 GHz, First Edition. Ralph Morrison.
© 2012 John Wiley & Sons, Inc. Published 2012 by John Wiley & Sons, Inc.

A	Ampere, the unit of current
ac	Alternating current. Usually a sine wave of voltage or current
A/D	Analog-to-digital
am	Amplitude modulation
amp	Ampere (not recommended)
ANSI	American National Standards Institute
ASCII	American standard code for information interchange
ASIC	Applications specific integrated circuit
ASSP	Application-specific standard product
ASTTL	Advanced Shottkey transistor–transistor logic
ATE	Automatic test equipment
B	Magnetic induction field, measured in teslas
BGA	Ball grid array
BH	B and H fields in a magnetic material. Hysteresis curves
BOM	Bill of materials
b/s	Bits per second
BST	Barium-strontium-titanate (ceramic)
BTL	Bipolar transistor logic
BTU	British thermal unit
BW	Bandwidth
c	Velocity of light
C	Capacitor, capacitance
C	Coulomb. The unit of charge
C_{12}	A mutual capacitance
CAD	Computer-aided design
CAM	Computer-aided manufacturing
cc	Cubic centimeter
CFFT	Complex fast Fourier transform
CISPER A and B	EMI standards that are set by the EU
cm	Centimeter
CMOS	Complimentary metal oxide semiconductor
cmr	Common-mode rejection
cmrr	Common-mode rejection ratio (analog)
cos	Cosine
cot	Cotangent
CPC	Computer power center
cps	Cycles per second. Not in common use
CRT	Cathode ray tube
CSA	Canadian Standards Association
csc	Cosecant
Cu	Copper
D	Displacement field
D/A	Digital to analog

dB	Decibel. $10 \log P_1/P_2$ or $20 \log V_1/V_2$. P_2 or V_2 are reference levels.
dBA	Decibel-amperes
dBm	Decibel-milliwatts
dBV	Decibel-volts
dBµV	Decibel-microvolts
dc	Direct current. A steady value describing volts, current, or field strength.
D_f	Dissipation factor
DFF	D flip flop
DFM	Design for manufacturability
DFT	1. Design for test
	2. Discrete Fourier transform
DIP	Dual inline package
D_k	Dielectric constant
DRAM	Dynamic random access memory
DTL	Decoupling transmission line
e	2.71828. Base of natural logarithms
E	Electric field strength. Basic unit is volts per meter
E	Energy in joules
ECAD	Electronic computer-aided design
ECL	Emitter-coupled logic
EDA	Electronic design automation
EE	Electrical engineer
EIA	Electronics Industry Association
EM	Electromagnetic
EMC	Electromagnetic compatibility
emf	Electromotive force, voltage
emi	Electromagnetic interference
ENIG	Electroless nickel/immersion gold
EPROM	Electrically programmable read only memory
ESD	Electrostatic discharge
ESL	Equivalent series inductance
ESR	Equivalent series resistance
EU	European Union
f	Frequency
F	Farad, the unit of capacitance
FB	Feedback
FCC	Federal Communications Commission
FET	Field effect transistor
FFT	Fast Fourier transforms
FIFO	First in first out
FILO	First in last out

fm	Frequency modulation
FP	Field programmable
FPGA	Field programmable gate array
FR-4	Flame resistant class 4 circuit board laminate. Glass epoxy
FTTL	Fast TTL
g	Gram
G	Gauss, 10^{-4} teslas
GaAs	Gallium arsenide
Gb/s	Gigabits per second
GHz	Gigahertz, 10^9 Hz
GND	Ground
GTL	Gunning transistor logic
H	1. Magnetic field strength, Basic unit is amperes per meter
	2. henry, unit of inductance
HASL	Hot air solder leveling
HDI	High density interconnect
HF	High frequency
HSTL	High speed transceiver logic
Hz	hertz (frequency). 1 Hz = 1 cycle per second
IBIS	I/O buffer information specification
i	1. Varying current.
	2. Square root of -1
IC	Integrated circuit
id	Inside diameter
IEEE	Institute of Electrical and Electronic Engineers
I/O	Input/output
IP	Intellectual property
IPC	Institute of Printed Circuits
IR	Infrared
ISDN	Integrated Services Digital Network
J	Joule
JEDEC	Joint Electronics Device Engineering Council
JFET	Junction FET
k	Kilo, 10^3
kb/s	Kilobytes per second
KB	Keyboard
kg	Kilogram
kHz	Kilohertz, 10^3 Hz
km	Kilometer, 10^3 m
kV	Kilovolt, 10^3 V
kW	Kilowatt, 10^3 W
kΩ	Kiloohms, 10^3 ohm
L	Inductor, inductance
L_{12}	Mutual inductance

LAN	Local area network
LED	Light emitting diode
LCC	Leaded chip carrier
LICA	Low inductance capacitor array
ln	Natural logarithm (base e)
log	Logarithm (base 10)
LRC	Inductor/resistor/capacitor
LVCMOS	Low voltage CMOS
LVDS	Low voltage differential signaling
m	1. Meter
	2. Milli
M	Mega, 10^6 (M mean 1000 in Roman numerals but not in engineering)
mA	Milliampere, 10^{-3} A
Mb/s	Mega bits per second
MCAD	Mechanical computer-aided design
MCM	Multi chip module
mH	Millihenry
MHz	Megahertz, 10^6 Hz
mil	0.001 inches
Mil	Military
Mil Std	Military standard
mJ	Millijoule
mm	Millimeter
mmf	Magneto motive force
MOS	Metal oxide semiconductor
MOSFET	Metal oxide semiconductor field effect transistor
ms or msec	Millisecond, 10^{-3} s
mV	Millivolt, 10^{-3} V
mW	Milliwatt, 10^{-3} W
mΩ	Milliohm, 10^{-3} ohm
MΩ	Megaohm, 10^6 ohm
N	Newton
n	Nano, 10^{-9}
nA	Nanoampere, 10^{-9} A
NEC	National Electrical Code
NEMA	National Electrical Manufacturing Association
nF	Nanofarad, 1000 pF, 0.001 µF, 10^{-9} F
nH	Nanohenry
nm	Nanometer
npn	Transistor made from a p-doped semiconductor layer between two n layers
ns	Nanosecond, 10^{-9} s
OC	Optical carrier

od	Outside diameter
OEM	Original equipment manufacturer
p	Pico, 10^{-12}
PC, pc	1. Printed circuit
	2. Personal computer
PCA	Printed circuit assembly
PCB	Printed circuit board
PCI	Personal computer interface
PE	Professional engineer
PECL	Positive ECL
p_f	Power factor
pF	Picofarad, 10^{-12} F
pH	Picohenry, 10^{-12} H
PLL	Phase-locked loop
pn	Junction of positive and negative doped semiconductor, diode
pnp	Transistor, n-doped semiconductor between two p layers
ps	Picosecond, 10^{-12} s
PTFE	Polytetrafluoroethylene (dielectric)
PWB	Printed wiring board
PWR	Power
Q	A measure of losses in resonance. Low loss equals high Q
QA	Quality assurance
QFP	Quad flat pack
R	Resistor, resistance
RAM	Random access memory
RC	Resistor/capacitor
rf	Radio frequency
rfi	Radio frequency interference, general interference
rms	Root mean square
RoHM	Reduction of hazardous material (lead free)
ROM	Read only memory
rpm	Revolutions per minute
rps	Revolutions per second
RTI	Referred to input
RTO	Referred to output
s	Second
SCR	Silicon controlled rectifier
sec	1. Second (not recommended)
	2. Secant
Si	Silicon
SI	Signal integrity
SIP	Single inline package
SMD	Surface mounted device
SMT	Surface mount transistor

SPICE	Special program for integrated circuit emulation
SRAM	Static random access memory
SWR	Standing wave ratio
t	Time
T	1. tesla, the unit of magnetic induction
	2. Temperature
tan	Tangent
TBD	To be determined
TC	Temperature coefficient
TDR	Time domain reflectometer
TE	Transverse electric (wave)
TEL	1. Transitional electrical length. The distance a wave travels in a rise time
	2. Telephone
TELCO	Telephone company
T_g	Resin transition temperature in laminates
TIA	Telecommunications Industry Association
TL	Transmission line
TM	Transverse magnetic (wave)
TQFP	Thin quad flat pack
TTL	Transistor–transistor logic
TV	Television
UHF	Ultrahigh frequency
UL	Underwriters Laboratories
USB	Universal serial bus
UTP	Unshielded twisted pair
v	Varying voltage
V	Volt
V	Volume usually in meters cubed
VA	Volt-amperes
VAR	Volt amperes reactive
Vcc	Positive voltage on a circuit board
VCR	Voltage controlled rectifier
Vdd	Positive voltage on a circuit board
VME	Type of bus and hardware protocol
VOH_{max}	Highest loaded voltage for a logic 1 of a driver
VOH_{min}	Lowest loaded voltage for a logic 1 of a driver
VOL_{max}	Highest loaded voltage for a logic 0 of a driver
Vss	The ground or most negative voltage on a circuit board
VSW	Voltage standing wave
W	Watt
W	Work
wan	Wide area network
X	Reactance in ohms

Y	Admittance in mhos
Z	Impedance in ohms
β	Current gain for a transistor
ε_R	Relative dielectric constant, permittivity
θ	Angle, phase, or phase angle
λ	Wavelength
μ_R	Relative permeability
μ	Micro, 10^{-6}
μA	Microampere, 10^{-6} A
μF	Microfarad, 10^{-6} F
μH	Microhenry
μm	Micrometer, 10^{-6} m
μs	Microsecond, 10^{-6} s
μV	Microvolt, 10^{-6} V
π	3.14159
ρ	1. Resistivity
	2. Reflection coefficient
σ	Conductivity
τ	Transmission coefficient
τ_r	Rise or fall time
ϕ	Angle, phase, or phase angle
ω	Radian frequency, $2\pi f$
Ω	ohm

BIBLIOGRAPHY

Archambeault BR. *Design for Real-World EMI Control*, Norwell, MA: Kluwer Academic Publishers, 2004.

Beaulieu D. *Printed Circuit Board Basics: An Introduction to the PCB Industry*, Canton, GA: UP Media Group, 2003.

Edwards TC. *Foundations for Microstrip Circuit Design*, New York, NY: John Wiley & Sons, Inc., Reprint 1987.

Howe H. Jr. *Stripline Circuit Design*, Norwood, MA: Artech House, Inc., 1974 Reprint on Demand.

Lawday G. *A Signal Integrity Engineer's Companion: Real-Time Test and Measurement and Design Simulation*, Upper Saddle River, NJ: Prentice Hall, 2008.

Morrison R. *Grounding and Shielding: Circuits and Interference*, Fifth Edition, Hoboken, NJ: John Wiley & Sons, Inc., 2007.

Pfeil C. *BGA Breakouts and Routing*, Wilsonville, OR: Mentor Graphics®, 2008.

Ramo S. *Fields and Waves in Modern Radio*, Second Edition, New York, NY: John Wiley & Sons, Inc., 1953.

Ritchey LW. *Right the First Time: A Practical Handbook on High-Speed PCB and System Design, Volumes I and II*, Glen Ellen, CA: Speeding Edge, 2003, 2007.

Wadell BC. *Transmission Line Design Handbook*, Norwood, MA: Artech House, Inc., 1991.

Digital Circuit Boards: Mach 1 GHz, First Edition. Ralph Morrison.
© 2012 John Wiley & Sons, Inc. Published 2012 by John Wiley & Sons, Inc.

INDEX

A/D converters, 98
Ampere's law, 13
Analog, 15, 66, 120
   carrier signals, 1
   definition, 1
   measure of characteristic impedance, 56
Antennas, 98
Apertures, 126
Asymmetric stripline, 135
Atoms, 5

B field, 13
Balanced,
   differential, 121
   odd mode, 25
   signals, 66
   transmission lines, 26, 51
Biot Savant, 13
Boards, 123
   decoupling, 92
   edge radiation, 26, 45, 78
   four layer, 143
   islands, 97
   layers, 97
   multilayer, 79
   resonances, 90
   six layer, 145
   two layer, 131
Bumps, 94

Cables, 61, 89
   shielded, 121
Capacitance,
   cross coupling, 10
   current flow, 9
   definition, 8

ground/power plane, 40
   mutual, 10
   parallel plates, 9
   self, 10
   work, 8
Capacitor, 8, 90
   decoupling, 46, 48
   four terminal, 87
   layers, 87
   natural frequency, 113
   stubs, 54, 92
   as transmission line, 39, 86
Carrier, 22
Centered stripline, 135
Characteristic impedance, 27, 28, 30, 62, 63, 85,
   94, 102
   accuracy, 115
   boards, 131
   measure, 56
   milliohm, 89
Charges,
   concentration of, 7
   electric, 5
   movement, 15
   surface, 6
   test, 6
Circuit boards, 61, 65, 132
   early, 2
Circuits,
   analysis, 4
   energy storage, 9
   field approach, 3
   theory, 62
Clock, 85, 91, 93, 95, 131
Coax, 89
Common mode, 66, 122

Components,
  capacitive, 71
  fields, 5
  ideal, 25
  inductive, 71
Conducted interference, 61
Conducting planes, 62
Conductivity, 95
Conformal coating, 28
Coulomb, 9
Coupling,
  far end 71
  forward wave, 71
  near end, 71
  odd mode, 69
  reverse wave, 72
Critical length, 2
Cross coupling, 10, 62 , 76, 77, 114, 145
Cross talk, 69, 71, 96, 100, 114
  inductive, 72
Current patterns, 6, 12
  low frequency, 19
Current in space, 12
  antennas, 12

D field, 12
Dc/dc converters, 124
Decoupling capacitor, 112, 135
Delays in transmission, 23, 84
Die capacitance, 112, 132
Dielectrics, 9 , 28
  losses, 97, 115
  thickness, 133
Differential,
  amplifiers, 66
  logic, 68
Digital signals, 5
Diode protection, 125
Dipole, 62
DIPs, 116
Displacement field, *see* D field
Drivers, 94
DTL, 87, 89, 114

E field, *see* electric field
E/H, 63
Earth ground, 119
Electric field, 5, 62
  dielectrics, 9
  field line termination, 7
  in space, 9
  penetration into conductors 7
  relation to D field, 12
  vector field, 6

Electrons, 5
  balance in atoms, 5
  forces, 5
  velocity, 8
Embedded stripline, 135
Energy dissipation, 79
Energy flow, 5, 27, 86, 88, 103
  board edge, 79
  capacitor, 8, 46
  density, 103
  fields, 4
  resonant circuit, 98
  through holes, 79
Energy management, 82, 94
Energy storage, 40
  capacitance, 12, 62
  E field, 13
  ground/power plane, 40
  inductance, 62
  transmission lines, 46
Equations, characteristic impedance, 132
Equipment ground, 118, 121, 124
Equipotential surface, 7
Error envelope, 107
Error list, 108, 114
ESD, 80, 125
ESL, 91
ESR, 91
Etching, 132
Even mode, 67

Facilities, 117
Farad, 9
Faraday's law, 13, 16
Fault protection, 118
Ferrite beads, 117
Field, *see* electric field and magnetic field
Field,
  dc, 19
  energy flow, 4, 8
  inside conductors, 6, 11
  intensities, 17, 63
  link between E and H, 14
Field lines,
  electric, 6
  magnetic, 13
Field pattern, traces, 23
Field solver, 112, 133
Floating conductors, 98, 121
Flooding copper, 98 , 131, 142
Flux, 126
Forces,
  electric fields, 5
  magnetic fields, 14

Fourier spectrum, 15
Forward wave, 72, 77
Four layer boards, 143
Frequency spectrum, 2

Gaps, 97
Gaskets, 126
Glass weave, 132
Ground, 2, 7, 66, 123
  earth, 119
  electrode system, 118
  equipment, 117
  facilities, 117
Ground bounce, 79
Ground pin, 142
Ground plane, 2, 7, 40, 84, 96, 98, 111
Ground/power planes, 98, 100, 110
  energy storage, 113
  time constant, 114
Ground split, 97

H field, 13, 62
Harmonics, 4, 97
Heat dissipation, 99
Holes, 131
Honeycombs, 126

IC, 84, 108
  decoupling, 93
Impedance, 4, 61
Inductance,
  definition, 13
  measure of, 14
  mutual, 16
  self, 16
Induction field, 13, *see* B field
Interference, 61, 65, 80, 98, 121, 124
Interposer board, 36, 94

Jitter, 69

Keyboards, 125

Laminates, 131, 144
Layers, 114
Layups, 146
Leading edge, 90
Lenz's law, 13
LICA, 93
Lightning, 80
Lines, *see* traces and transmission lines
Logic signals, 15, 93
  balanced, 51
  delays, 52

error list, 109
  voltage drops, 110
Loops, 62, 80, 86, 122, 142
Loss tangent, 97
Lumped parameter lines, 23, 71

Magnetic field, 13, 14, 16
  coupling, 72
  energy storage, 14
  in space, 14
Magnetic flux, 13, 16, 126
Manhattan pattern, 142
Maxwell's equation, 62, 95, 133
Measurements, 115
Microstrip, 26, 71, 133
Microwaves termination, 24
Mutual capacitance, 10
  measurement, 11

Natural frequency, *see* resonance
Near field/far field, 63
NEC, 120
Net lists, 107
  performance, 111
Networks,
  capacitor, 93
  energy, 54, 88
  program, 55
  resistor, 116
Noise budget, 61, 84, 108

Odd mode, 69
Open circuit transmission line, 31
Optical coupler, 121

Parallel terminations, 30
  locations, 116
Permeability, 13
Permittivity, 9, 12
Pinouts, 131
Plane waves, 63
Potential differences, *see* voltage
Power flow, 17, 84
Power line
  interference, 124
  filters, , 124
Power plane, 2, 40, 84, *see* ground/power plane
Power supply ripple, 114
Power time constant, 84
Power transistors, 102
Poynting's vector, 17, 31
Prepreg, 143
Press fit, 145
Probes, 79, 111

Propagation of waves, 28
Protons, 5
Pulse frequency, 63

Radiation, 18, 61, 98, 99
    board edges, 78
    circuit boards, 65
    ground split, 97
    microstrip, 70
    transmission lines, 19, 30, 78
Reference conductor, 10
Reflection, 91, 132
Reflection coefficient, 35
Relative dielectric constant, 9
Relative permeability, 13
Reverse wave, 72, 77
Resistance under trace, 111
Resistivity, 110
    copper, 111
Resistors, 69, 132
    accuracy 115
    termination, 24
Resonances, 90
    capacitors, 90
    parallel, 91
    series, 91
Resonant circuit, 2, 90
Return current, 27, 144
Return path, 96, 120
rf, 22
Ripple, 114
Rise and fall time, 15, 52, 95
Rise time rule, 70

SCR, 125
Separately derived power, 119
Series termination, *see* transmission line
Series termination, 70, 77
Shield, 98, 145
    traces, 78
Shielding, 125
Shockley junction, 37
Short circuit transmission lines, 32
Shunt terminations , 30
Signal,
    common, 120
    delay, 2
    integrity, 76, 106, 131
    reference, 120
Simulations, 106
SIPs, 116
Six layer boards, 145
Skin effect, 8, 11, 91, 95
Solder, holes, 13

Source impedance, 52, 88
Source terminations, 30, *see* series terminations
Specifications, 106
Spectrum, 2
Square wave,
    harmonic amplitudes, 15
    spectrum, 15
Steady state, 2
Step function, 22, 61, 63, 90
Step voltage, 10
Step waves, 28, 30, 90
Stripline, 51, 71, 91
Stubs, 52, 69, 79, 90, 101, 131
Susceptibility, 80
Switch, 70, 93
    ideal, 24

TEL, 77
Terminations, 69, 132
    capacitance, 112
    non-linear, 37
    parallel or shunt, 30, 50
    series or source, 49, 70
    stripline, 25
    transmission lines, 25
Thieving, 145
Time constant, 44, 85, 114
Time delays, 2
Traces, 62, 84, 101
    count, 131, 142
    guard, 78
    low Z, 101
    resistance, 95
    through planes, 100
    two layer, 141
Transfer impedance, 122
Transformer, 77
Transients, 2
Transmission coefficient, 35
Transmission lines, 18, 39, 66, 82, 84, 86, 88, 93,
        113,
    balanced, 2
    cascaded, 46
    coaxial, 26
    conformal coating, 2
    crosstalk, 26
    current patterns, 26
    dielectrics, 28
    discharge, 39
    field pattern, 23, 26
    heat loss, 18
    interference, 26
    microstrip, 26
    open circuit, 31, 38

parallel termination, 30
propagation, 22
reflection coefficient, 35
series terminations, 30, 49, 89, 92
stripline, 27
tapered, 41
terminations, 25
time constant, 39
transmission coefficient, 35
uses, 18
voltage doubling, 31
TTL, 41, 107
Two layer boards, 131, 135
Two-ounce copper, 110

UL, 120

Vacuum tubes, 130
Vector field, *see* electric field and magnetic field
Vias, 27, 36, 78, 86, 100
extensions, 132
pattern, 143
Voltage, 5
definition, 6
doubling, 31

induced, 15, 125
logic limits, 107
Voltage source,
ideal, 24, 85
step, 24

Water analogy, 87
Wave guide, 126
Wave impedance, 62
Waves, 28, 38, 61, 63, 72, 78, 84, 86
forward, 72
impedance, 63
leading edge velocity, 29
plane, 63
reflected, 30
reverse, 72
tapered transmission lines, 41
Weave, 132
Wirewrap, 130
Work
electric field, 8
magnetic field, 14

Zapper, 125

Printed and bound by CPI Group (UK) Ltd, Croydon, CR0 4YY

27/10/2024

14580268-0005